Hickman's Analog and RF Circuits

Hickman's Analog and RF Circuits

Ian Hickman
BSc (Hons), C. Eng., MIEE, MIEEE

Newnes
An imprint of Butterworth-Heinemann
Linacre House, Jordan Hill, Oxford OX2 8DP
A division of Reed Educational and Professional Publishing Ltd

℞ A member of the Reed Elsevier plc group

OXFORD   BOSTON   JOHANNESBURG
MELBOURNE   NEW DELHI   SINGAPORE

First published 1998

British Library Cataloguing in Publication Data
A catalogue record for this book is available from the British
Library.

ISBN 0 7506 3742 0

Library of Congress Cataloging in Publication Data
A catalogue record for this book is available from the Library of
Congress.

Typeset by Vision Typesetting
Printed and bound by Antony Rowe Ltd, Eastbourne
Transferred to digital print on demand, 2005

Contents

Preface

I have been writing articles for magazines since the early 1970s, and many of these have been published in *Electronics World* (founded in 1911 as *The Marconigraph* but known for most of its life as *Wireless World*). This is undoubtedly the foremost electronics magazine in the UK, being widely read by professional electronics engineers on the one hand, and the more advanced electronics enthusiasts and hobbyists on the other, both in the UK and throughout many countries of the world, English-speaking and otherwise. A collection of these articles was published in 1995, under the title *Analog Circuits Cookbook*. This proved very popular, and the present companion volume is the result of a suggestion that further articles of mine from *Electronics World* should likewise be republished. Most but not all in this present collection of articles were published since the preparation of the earlier volume. And all appeared under the name Ian Hickman except two, which originally appeared under other pen names.

Inevitably, in the preparation for publication of a magazine which appears every month, the occasional 'typo' crept into the articles as published, whilst the need to adjust articles to fit the space available led to the occasional pruning of text. As in the previous volume, the opportunity has been taken to restore the excised material, whilst it is hoped that most if not all errors in the articles as published have been identified and corrected. Each article has been prefaced with a brief introduction, indicating the contents and its general drift.

By and large, each of the articles deals with one or other of my two main areas of expertise – general analog electronics, and RF. But a significant minority deal with the theory behind the various practical areas of circuit design or operation. Explanatory articles of this sort have long been a feature of *Electronics World* and the former *Wireless World*. For many years, such articles appeared from time to time from the pen of one 'Cathode

Ray', who had the knack of explaining things clearly, without resort to extensive mathematical treatment. These articles were always much appreciated by readers, and it is an open secret that the author was none other than that well-known writer on electronics, M G Scroggie. To some extent, it seems that the mantle of Cathode Ray has latterly fallen on my shoulders; at least, I hope that the articles in this volume dealing with the fundamentals of electronics will prove both interesting and easily assimilable.

The articles have been grouped in four sections: General analog circuitry, Audio, RF and Basic principles. The latter section contains the 'Cathode Ray'-style explanatory articles, though it must be said that to some extent any classification scheme must be arbitrary – some articles could equally well be included under two if not three of the four groupings mentioned above. But that is neither here nor there; the important things are the articles themselves. So now, please read on . . . .

Ian Hickman, 1997

Part 1
GENERAL ANALOG CIRCUITRY

1 Voltage references

Some of the devices mentioned in this article, which appeared as long ago as 1991, may by now be scarce or unobtainable. But the circuit considerations discussed are important and valuable, so the piece is well worth including. Modern equivalents of the devices mentioned – often with improved performance – are in any case available.

Voltage references have made great strides in the last quarter of a century, since solid state types replaced gas tubes. First on the scene was the zener diode, used with a simple series current-limiting resistor, Figure 1.1(a). Popular ranges were the BZY88 series (Mullard/Philips) and the American 1N821-827 series. The slope (incremental) resistance and the temperature coefficient of a zener diode each affect both the stabilisation (change of output voltage with change of supply voltage) and the regulation (change of output voltage with output current drawn) of the circuit. The operating mechanism is true zener breakdown for low voltage diodes and avalanche breakdown for higher voltage types. The former has a negative 'tempco' and the latter a positive. Accordingly at the changeover point – somewhere between 4.7 and 5.6V where both mechanisms occur – the tempco is near zero, varying with the device process used and also somewhat with operating current. Thus it was tempting to use a nominal 5.1V diode where high stability was required, but unfortunately the lowest slope resistance is

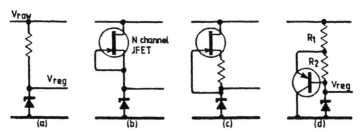

Figure 1.1 *Voltage references built around the zener diode.*

found in diodes of about 7.5V nominal; these were therefore often preferred where good regulation and particularly stabilisation were important. However, the quoted slope resistance of a zener is always measured adiabatically, by superimposing a small AC ripple component on the DC current and observing the ripple voltage. The temperature of the diode does not vary in sympathy with the instantaneous current as the frequency of the AC is too high, so the tempco does not influence the result. Many an unwary circuit designer has chosen a zener with the lowest slope resistance only to find the apparent stabilisation of his circuit much worse than expected. When the supply voltage is suddenly increased, the increased dissipation in the diode raises its temperature and the positive tempco then contributes a rise in output voltage above and beyond that due to the extra current flowing through the slope resistance. In Figure 1.1 (b) and (c) the high slope resistance of a depletion mode FET greatly reduces any current variations through the diode due to change of supply voltage. But a FET is comparatively expensive, and anyway has an embarrassingly high variation, up to 5 : 1, in drain saturation current I_{dss}. This is controlled to some extent by the self-bias resistor in Figure 1.1 (c). The circuit of Figure 1.1 (d) seems to be comparatively little known but is very economical. Any increase in current through the zener diode due to increased supply voltage would tend to increase the base emitter voltage across R_2, shunting most of any increase in current through R_1 to ground via the collector circuit.

Straightforward zeners had a long run for their money, with many ingenious circuit variations to improve regulation and stabilisation. One I developed many years ago, published in *Wireless World* under the title 'Two for the Price of One', provided two stabilised supply rails using a single zener diode. However, my favourite augmented zener circuit is shown in Figure 1.2(a). Here, the zener is incorporated in a bridge circuit which is driven by an op-amp whose input signal is the bridge output. The circuit is thus a little incestuous and usually needs a resistor (shown dotted) to ensure reliable start-up. However, its value can be as high as 10MΩ, so that although it feeds some current into the zener from the unregulated input voltage, its effect on stabilisation is negligible. The op-amp buffers the load current, resulting in excellent regulation, whilst the bridge is fed from a constant voltage, resulting in constant current through the zener and hence excellent stabilisation. This circuit is available in IC form, for example the Burr-Brown REF10, which is shown in Figure 1.2(b) and provides 1ppm/ °C max. tempco, 10ppm/1000hrs stability, 6μV p/p noise, 0.001%/V stabilisation and 0.001%/mA regulation. The output voltage is 10V ± 5mV with facility for trimming to 10V exactly.

The on-chip zener diode used in the REF10 runs at 6.3V and provides a near zero tempco with the IC process used. However, another very popular IC form of reference voltage is based on the band-gap principle, where two transistors are run at different current densities. Many manufac-

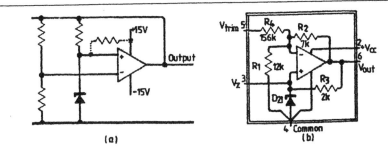

Figure 1.2 *Two of the many augmented zener circuits developed.*

turers offer this type of device, a typical example being the Analog Devices AD580. This is a three terminal device (V_{supply}, V_{out} and ground) supplying a $2.5V \pm 0.4\%$ output with a 1Ω output resistance, $25\mu V$/month stability and a low quiescent current of $1.5mA$. Another range of bandgap references are the REF12, 25, 50 series of $1.26V$, $2.5V$, and $5.0V$ devices from Plessey Semiconductors. These are two terminal devices, needing an external current limiting resistor just like a zener diode. However, the 'knee current', that is, the current at which the slope resistance has fallen to a low value, is very low for these devices – less than $100\mu A$, making them ideal for battery-powered applications. They are also available in trimmable, three-terminal styles and in SMD – surface mount – packages.

Voltage references are available in a wide range of output voltages such as $2.5V$, $5.0V$ and $10.0V$ whilst others cater for binary-oriented instrumentation, with $2.56V$, $10.24V$ outputs, etc. Two-terminal types are subject to a selection tolerance but many three or more terminal devices offer the facility to adjust the output voltage exactly to the nominal. Where an output voltage is required which is not available as standard, the circuit of Figure 1.2(a) is ideal; by choosing appropriate values for the bridge resistors, any desired output voltage can be obtained. Where several different voltages are required simultaneously, the highest can be produced with a Figure 1.2(a)-type circuit and the others obtained from it with a potential divider string of precision (or adjustable) resistors, the tappings being buffered with non-inverting op-amps.

Often, however, only one voltage is required at a time, but it must be adjustable to any one of a number of possibilities. A typical application might be in ATE. In this case, the above schemes are less attractive and another arrangement must be sought. A simple solution, particularly attractive in ATE where a microcontroller is incorporated to organise the operation of the system, is to use a voltage reference and a multiplying DAC, as indicated in Figure 1.3(a). The DAC chosen will determine the resolution to which the voltage can be set. For instance, with an 8-bit DAC and a $10.24V$ reference such as National Semiconductor's LH0071-OH, the resolution will be $40mV$, i.e. 0.39% of full scale, which in many cases

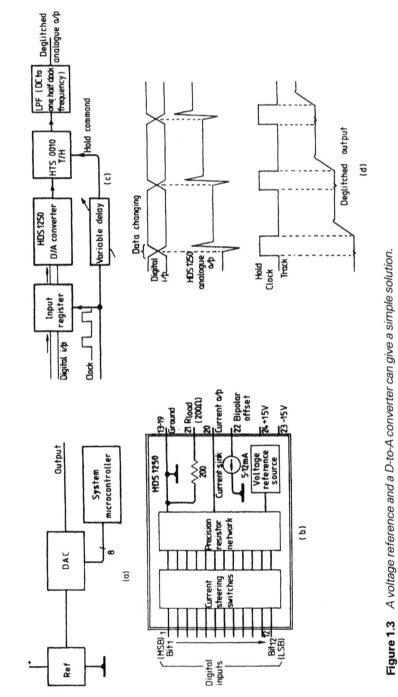

Figure 1.3 *A voltage reference and a D-to-A converter can give a simple solution.*

would be inadequate. Using a 10-bit DAC, such as Philips' MC3410F, would improve the resolution to 10mV or better than 0.1% FS, whilst a 12-, 14- or 16-bit multiplying DAC would clearly provide much greater resolution. The accuracy of the output would also improve generally in line with the resolution, but the designer must decide just what he needs and can afford: many DACs are available in various selection grades offering an accuracy which may be as high as one quarter of an LSB (least significant bit) in the better devices to as poor as one and a half LSBs in the more economical sort.

Instead of using a reference source and a separate DAC as in Figure 1.3(a) an attractive option is to use a multiplying DAC with a built-in reference. The Analog Devices HDS-1250 (Figure 1.3(b)) is a high-performance example, being a 12-bit device with the very fast settling time of 35ns. It can be used in either current- or voltage-output mode, in the former providing a full-scale output of 10.24mA – or more strictly speaking 10.2375mA with an all-ones input code (10.24mA would correspond to an all-noughts input code plus a one in the non-existent thirteenth bit). Glitches commonly occur when the output of a DAC changes, and these can be pronounced if there is any skew between the various bits of the control word. In some circuit applications, this can be most embarrassing. One way to reduce these is to include input buffer registers in the DAC, as in the AD7224 CMOS monolithic 8-bit double-buffered voltage output DAC from MAXIM. In addition to effectively de-skewing the input data word, double buffering permits simultaneous updating in systems where several DAC channels are successively written to by a single controller. Figure 1.3(c) shows alternatively how a T/H (track-and-hold circuit) can be used in conjunction with an unbuffered device such as the HDS-1250 mentioned earlier. The minor residual switching transients introduced by the action of the T/H are very fast and contain little energy, Figure 1.3(d): they are easily eliminated by a little light low-pass filtering.

In electronics there are usually many ways of achieving the desired result and that is true here, especially where only a few output levels are required. In this case a very economical solution is to use a voltage reference plus a mux-amp. The latter is an op-amp with two or more input stages, only one of which is activated at a time. The gain is thus determined by the feedback components associated with the selected stage. A good example is the National Semiconductors LM604, which has four input stages and a typical application is shown in Figure 1.4, where two data bits determine, under control of the CHIP SELECT BAR and WRITE BAR pins, which of four output levels is selected. This versatile chip also includes a facility to disable the output, so that the outputs of two such chips may be paralleled. Thus it is possible to extend the scheme to eight output levels by using a third bit and its logic inverse to select one or other LM604 via the ENABLE BAR inputs.

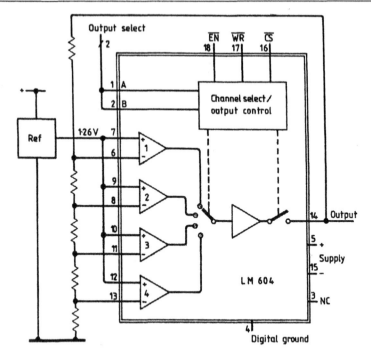

Figure 1.4   *Circuit providing one of four different outputs of 1.26V or more.*

2 Versatile twin amplifier has many uses

OTAs – operational transconductance amplifiers – are exceedingly versatile analog building blocks, with a host of possible applications. The utility of an OTA is further enhanced when, as here, it is combined with a wide bandwidth buffer stage.

There are many dual op-amps available, but the subject of this design brief is not a dual, but rather a *twin* amplifier. The LT1228 from Linear Technology Corporation contains an operational transconductance amplifier (OTA) with a maximum bandwidth of 75MHz and a current feedback amplifier (CFA) with a 100MHz bandwidth. In the interests of packaging all this in an 8-pin plastic dual-in-line, the single-ended current output of the transconductance amplifier is tied internally to the NI (non-inverting) input of the CFA, which may be used as a buffer. This junction is also brought out to a pin and since the NI input resistance of the CFA is very high (25MΩ typical), it may be ignored and the OTA used on its own if desired. The inclusion of a buffer is, however, a great convenience to the designer in many applications and is found in some earlier types of OTA such as the LM13600/13700 (but not others such as the CA3060/3080), though the frequency range of these earlier devices was much more limited. One of the more obvious applications for an OTA is as an electronically-controlled variable gain stage, and Figure 2.1(a) shows such a circuit with an input resistance of 10kΩ, a gain range of -18 to $+2$dB and a -3dB bandwidth of around 20MHz. The input may be differential as shown, or unbalanced, inverting or non-inverting, in which case R_{3A} or R_{2A} respectively may be omitted. The gain is directly proportional to I_{set}, the current into pin 5 of the device. The Thévenin source arrangement shown with R_4 and R_6 compensates for two internal diode drops in the gain-setting section and will thus ensure (for stabilised 15V rails) that any

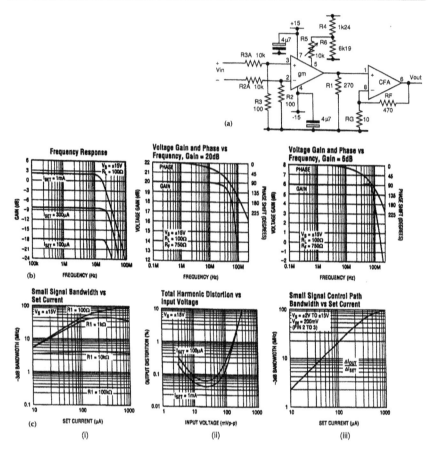

Figure 2.1 *(a) Variable gain amplifier with bandwidth of around 20MHz and gain adjustable over the range −18 to +2dB. (b) Variation of gain of the CFA with frequency, for three different values of demanded gain. (c)(i) Bandwidth of the OTA section; (ii) THD of the OTA section as a function of input signal amplitude; (iii) small signal control path bandwidth of the OTA section versus set current I_{set}.*

set gain remains constant within 1% over the device's full temperature range of −55 to +125 °C (note that R_4 returns to the negative rail). Other resistor values are appropriate if using a different negative supply rail voltage, whilst if this rail is not stabilised, compensation may be achieved using an LT1004 −2.5V reference. Alternatively, for more accurate and linear control of gain, I_{set} may be supplied by a single op-amp voltage-to-current converter circuit. The input attenuator ensures that the circuit can accept inputs up to 10V pk-pk (1k24 connects to −15V).

The mutual conductance g_m (output current divided by the voltage between pins 2 and 3, in mA/V) is given by $g_m = 10 \times I_{set}$, and in the circuit shown this flows in R_1, buffered by the high input impedance of the CFA. The voltage gain of the latter is equal to $(R_f + R_g)/R_g$ at low frequencies, and up to the frequency at which the CFA's gain bandwidth product of about 1GHz becomes significant. (How the bandwidth of the CFA varies with demanded gain is shown in Figure 2.1(b). Thus overall, the gain A_v in Figure 2.1(a) is given by $A_v = R_3/(R_3 + R_{3A}) \times 10 \times I_{set} \times R_1 \times (R_f + R_g)/R_g$. If the maximum input expected is less than 10V pk-pk, the 10K resistor(s) at the input may be reduced, giving an increased A_v. If an increase in A_v is not required, then R_g may be increased, demanding less gain from the CFA and increasing the circuit's bandwidth. But any substantial increase in bandwidth may be limited by the bandwidth of the OTA section, which is shown in Figure 2.1(c)(i). Figure 2.1(c)(ii) shows the THD (total harmonic distortion) of the OTA section as a function of input signal amplitude. In the application in Figure 2.1(a), I_{set} is basically a DC whose value is adjustable for any desired gain. However, in some applications (e.g. Figure 2.6 on p. 17) high frequency signals may be applied to the control path input at pin 5, and Figure 2.1(c)(iii) shows the small signal *control path* bandwidth versus set current I_{set}.

Another major application for OTAs is in electronically tuned filters. A single-pole filter is the simplest possible type, offering a flat pass-band with a -6dB/octave roll-off in the stop-band beyond the -3dB corner frequency. The latter can be electronically controlled over a wide range, Figure 2.2 being an illustration of a suitable circuit arrangement. For operation as a low-pass filter the high-pass input should be grounded, and vice versa. Considering the low-pass case, at high frequencies where C is almost a short circuit, there is little output and what there is will be in quadrature. On the other hand, at low frequencies, where C is effectively open circuit, the voltage gain of the OTA is indefinitely large, and is

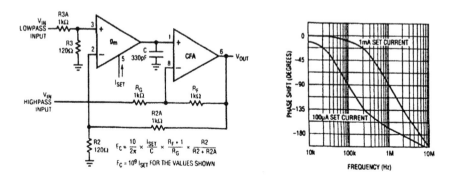

Figure 2.2 *Single-pole low-high-all-pass filter. For low-pass operation the high-pass input should be grounded and vice versa. With the two inputs tied together, an all-pass response is obtained.*

included along with the non-inverting gain of two of the CFA (the HP input is grounded) within an overall NFB loop to the OTA's inverting input, pin 2. As the two voltage dividers at the OTA's inputs have the same ratio, there is unity non-inverting gain from low-pass input to output. The -3dB point where the phase shift through the circuit is $-45°$ can be set by adjusting the current I_{set} into pin 5. If the low-pass input is grounded instead, a high-pass response is obtained, with the same -3dB corner frequency and unity *inverting* gain in the pass-band.

With the two inputs tied together, an all-pass response is obtained, as predicted by the Theorem of Superposition, passing from zero phase shift at 0Hz through 90° at the corner frequency to 180° at high frequencies. Two LT1228s can be configured to give electronically tunable versions of any of the standard second-order filter sections, and the data sheet includes the ingenious circuit shown in Figure 2.3, which accepts inputs up to 3V pk-pk. This version of the state variable filter does not need a third inverting op-amp (as required when using conventional integrators), since the OTA integrators have both inverting and non-inverting inputs available, but if one were used, then a high-pass output would also be available. The circuit provides the novel feature of logarithmic tuning sensitivity and could thus be turned into a logarithmic sweep generator, by raising the value of the damping resistor R_d, connecting antiparallel diodes in series with it, and ensuring oscillation by including some negative damping also to the non-inverting input, pin 3, of the upper OTA.

However, all-pass circuits can also be configured as an oscillator. The first such example probably predates World War II and several such designs having appeared in this very journal. One of these,[1] was a very low distortion audio oscillator covering 20Hz to 20kHz and using an ingenious distortion out-phasing scheme. The earliest such all-pass design I know of was called 'The Selectoject' and operated as notch- or narrow band-pass filter or oscillator as required, but a trawl through my file of articles saved from *Wireless World* over the years has failed to unearth it. Figure 2.4(a) shows the circuit of an all-pass oscillator with which I experimented. An initial ill-judged attempt to build the circuit on experimenter's plugboard was predictably unsuccessful, so it was rebuilt using single-sided copper board as a groundplane. Since both of the all-pass stages are non-inverting at DC, a third LT1228 was employed to give the necessary inversion, providing overall NFB and hence stability at 0Hz: it was also used to stabilise the amplitude of oscillation. Figure 2.4(b) shows the gain and phase of the circuit with the loop broken, but with the I_{set} applied to pin 5 of IC$_3$ equal to what it is when the loop is closed. At the corner frequency of the two all-pass stages, each contributes 90° phase shift, giving a total loop gain of exactly unity, non-inverting, and hence stable oscillation. This occurs at a level which just turns on Tr1 on positive-going peaks, reducing the I_{set} available to IC$_3$ as necessary. Figure 2.4(c) shows the output

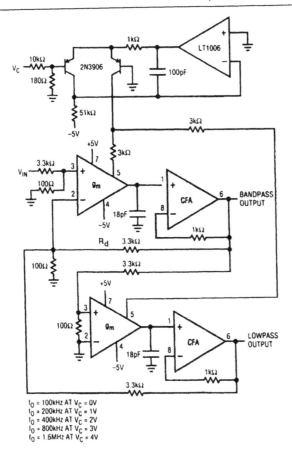

Figure 2.3 *Second-order electronically tuned state variable filter with low-pass and band-pass outputs. Features logarithmic tuning sensitivity.*

waveforms of IC_3 (leading trace) and IC_1 when the tuning control RV_1 is set for a 2MHz output, showing the low distortion and accurate quadrature. The circuit operates from well below 1MHz to beyond 5MHz, but by 5MHz the quadrature phasing is less than $90°$, due to the onset of additional loop phase shift in the inverting stage IC_3. Beyond about 7MHz, this becomes so marked that the circuit switches to a different mode of oscillation with around $60°$ phase shift in each of the three stages, operating thus up to 25MHz or more. Figure 2.4(d) shows the spectrum of the output of IC_2 at 2MHz (2MHz/div. horizontal, start $= 0$Hz); at 1MHz and below all harmonics are more than 40dB down.

The OTA is a versatile building block, enabling amongst other things an electronically controlled resistor to be simulated, by grounding its non-inverting input and connecting its inverting input to its output. Thus if the

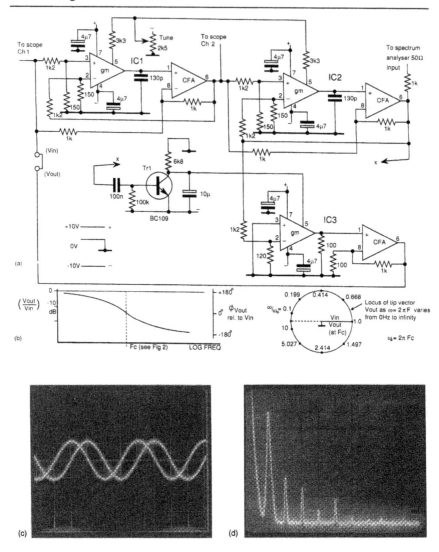

Figure 2.4 *(a) Circuit diagram of an all-pass filter-based oscillator. The 0.6V Vbe of Tr1 stabilises the amplitude at 1.2V pk-pk, by robbing I_{set} from IC_3 until the loop gain just equals unity. (b) Open loop gain and phase shown as Bode and vector plots. (c) Output waveforms of IC_1 and IC_3, showing low distortion and accurate quadrature (horizontal 125ns/div.; vertical 500mV/div.). (d) Spectrum of output from IC2, showing second harmonic content 38dB below the fundamental 2MHz output and all other harmonics greater than 40dB down (horizontal 2MHz/div.; vertical 10dB/div.).*

output is taken positive with respect to ground, the OTA will sink current and will source it if taken negative, just as a resistor would. Figure 2.5(a) shows this arrangement used as part of an attenuator, to stabilise the amplitude of the output from a single supply spot frequency Wien bridge oscillator. The gain of ×34 supplied by the CFA keeps the swing at the input to the OTA down to 15mV, in order to avoid distortion due to overdrive. This precaution is necessary since for lowest distortion the LT1228, like all OTAs, can only accept a limited input swing; THD reaching 0.2% at 30mV rms input. An OTA's permissible input voltage swing is limited because there is no emitter-to-emitter degeneration in the input stage (as is clear from Figure 2.5(b)), and OTAs are frequently

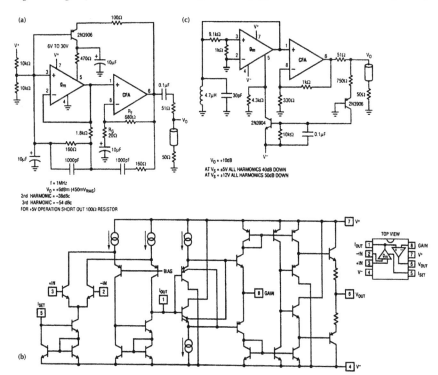

Figure 2.5 (a) Fixed frequency Wien Bridge oscillator using the LT1228. The oscillation is maintained by the CFA; the OTA functions only as an electronically controlled resistor forming an attenuator, in conjunction with the 1K8 resistor, to stabilise the amplitude of oscillation. (b) Simplified diagram of the internal workings of the LT1228 operational transconductance amplifier plus current feedback amplifier. (c) The OTA forms an electronically controlled negative resistance connected across an RF tank circuit, cancelling the losses and maintaining a constant level of oscillation.

required to operate with no overall feedback to constrain the inverting to non-inverting input voltage to a small value. Grounding an OTA's inverting input and connecting its non-inverting input to its output also simulates a resistance, a negative one in this case. Figure 2.5(c) shows such a negative resistor connected across an RF tank circuit, so as to cancel the losses and raise the tuned circuit's dynamic resistance R_d to infinity. Here, the 9K1/1K network at the OTA's input keeps the drive to a level that the device can handle linearly. Again, a transistor is used as a detector to sense the amplitude of the output from the CFA buffer and adjust the I_{set} of the OTA to stabilise the amplitude of oscillation. An intriguing possibility is the use of this circuit to maintain a constant very low level of oscillation in the tuned circuit of a simple radio receiver over its entire tuning range, acting both as an automated reaction control and as AGC. Such a receiver would handle both AM and SSB signals, offering very good selectivity, due to the tuned circuit operating at a very high Q.

The OTA has yet another party trick up its sleeve: it can function as a squarer, and hence as a frequency doubler. The current swing at the output of an OTA is proportional to the amplitude of the signal applied to the inverting or non-inverting input and also to the magnitude of I_{set}, i.e. to the product of those two quantities. If therefore a signal is applied simultaneously to both the inverting input and the I_{set} input, the output current will contain a component representing the square of the input voltage. The resultant circuit is only a two quadrant multiplier – the signal input can be bipolar but the I_{set} current must always be greater than zero, or the device simply cuts off. So the input merely modulates the (always positive) magnitude of I_{set}, the DC component of which is responsible for a component in the output current corresponding to the original input. To try the scheme out, I quickly knocked up a doubler circuit using the LT1228 on the lines I have described, but with a crucial addition, Figure 2.6. Since the signal is applied to the *inverting* input, pin 2, all components of the voltage developed across the 100Ω load resistor at the output are inverted in phase relative to the input. The component of the input voltage in the output can thus be cancelled (out-phased) by adding in a component from the input via the upper 1K2 resistor, leaving just the squared component. The square of sin(wt) being $(1 - \sin(2wt))/2$, the circuit will thus double the frequency of an input sinewave to 2wt radians per second, with – due to the out-phasing – no component at the original wt. The circuit is purely aperiodic, no frequency conscious components are involved (other than the trimmer at pin 2, which extends the operating frequency range of the circuit by compensating for slight phase shift in the OTA at higher frequencies). The output will therefore be a pure sinewave of twice the frequency of the input (assumed a sinewave) over a wide range of frequencies, though it would be wise not to rely on as much suppression of the fundamental input as seen in Figure 2.6(c).

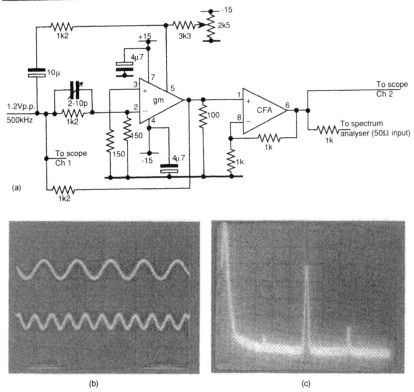

(a)

(b)   (c)

Figure 2.6 *(a) Circuit using an LT1228 as a four quadrant multiplier by out-phasing the residual component of the input waveform in the output. (b) The circuit of (a) doubling a 500kHz sinewave to 1MHz. Upper trace, input, 1V/div.; lower trace, output, 50mV/div.; horizontal 1µs/div. (c) Spectrum of output of (a). The 500kHz input is suppressed by 40dB and the third harmonic of the input at 1.5MHz is 35dB down relative to the wanted doubled output at 1MHz. All other outputs were at or below analyser noise level.*

As a final example of the countless applications for this versatile part, Figure 2.7 shows the circuit of a video crossfader. This uses two LT1228s in the feedback loop of an LT1223 CFA. Each of the two video inputs is applied via a 1K resistor to the OTA section of an LT1228, the CFA sections being unused. The two OTA output currents are connected to the *inverting* input of another CFA, the inverting input of a CFA being the low-impedance current-driven input. Negative feedback is applied from the output of the CFA to the non-inverting input of each OTA, via a 1K resistor, giving unity gain to each signal when the wiper of the 10K potentiometer is at mid-travel. The amount of signal from each input at the output is set by the ratio of the set currents of the two LT1228s, not by

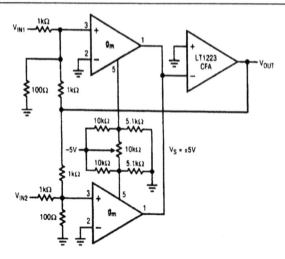

Figure 2.7   *Circuit of a video fader. Relative to unity gain (pot at mid-travel) a 15MHz bandwidth to each signal is maintained down to an attenuation of 20dB, with complete suppression of one or other signal at each end of the pot travel.*

their absolute value. The set currents remain high over most of the pot's range, keeping the bandwidth to each signal in excess of 15MHz, even when it is attenuated by 20dB. By this time, the other signal is dominant in the output video, and as the pot reaches the end of its travel, the attenuated signal is turned off completely.

Acknowledgements

Several of the illustrations in this article are reproduced by courtesy of Linear Technology Corporation.

Reference

1. Rosens, R. 'Phase-shifting oscillator', *Wireless World*, Feb. 1982 pp 38–41.

3 Differentiating op-amps

When choosing an op-amp for a particular application, it is a case of 'horses for courses' – depending on the particular performance characteristics most critical for a given application. This article compares two op-amps, each of which offers premium performance in some of its characteristics, at the expense of more modest performance in others. The two devices are as different as chalk and cheese.

It is a commonplace that in an ideal op-amp, bandwidth, open loop gain, differential and common mode input impedance, CMRR and PSRR would all be infinite whilst input offset voltage, input bias current and output impedance would all be zero – and surely there are one or two other desirables that I have overlooked as well. In practice, for any given application one must pick an op-amp that excels in those particular parameters which are essential, whilst accepting that the others will be less so. The following contrasts two bipolar op-amps with temperature coefficients of $V_{input\ offset}$ in the microvolt range, but which are otherwise as different as chalk and cheese, making each fitted to a particular task which would be beyond the capabilities of the other.

The CLC425 (Comlinear Corporation) op-amp features a 1.7GHz gain bandwith product, very low input noise ($1.05\mathrm{nV}/\sqrt{\mathrm{Hz}}$, $1.6\mathrm{pA}/\sqrt{\mathrm{Hz}}$ though the 1/f corner frequency is around 500Hz), a typical input offset voltage of 100μV, and a 360V/μs slew rate. Performance figures which fit it for ultrasound and instrumentation sense amplifiers, magnetic tape and disk preamps and a host of other applications. These include RF, in view of its bandwidth, which at a non-inverting gain of ×10 extends to 350MHz at ±1dB, Figure 3.1(a), fitting it for RF instrumentation applications. In fact, whilst its AC performance is reminiscent of a current feedback op-amp, being in fact a voltage feedback type, it provides vastly improved DC characteristics, such as a 2μV/°C input offset-voltage tempco. This amplifier is a 'decompensated' type, that is to say that whilst not *uncompensated*, it is not as heavily internally compensated as would be necessary for use at

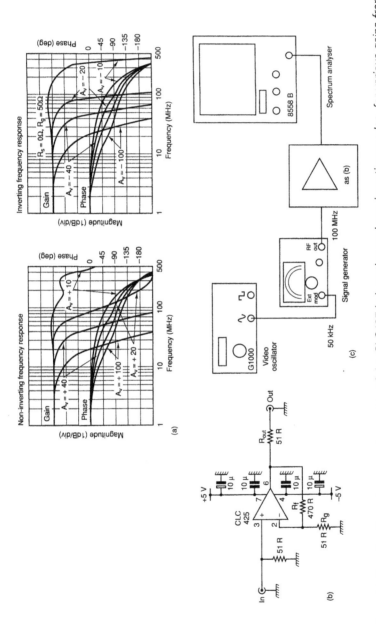

Figure 3.1 (a) Showing the gain versus frequency of the CLC425 in inverting and non-inverting modes, for various gains from ×10 upwards. (Reproduced by courtesy of Comlinear Corporation.) (b) Circuit used for evaluation of the CLC425 ultra low noise wideband bipolar op-amp. Internal to the board, the device produces a voltage gain of ×10, but as the input is terminated in 51R and a 51R source resistance R_{out} included in series with the output, the insertion gain is ×5 or just 14dB. (c) Test set-up used for the subsequent measurements.

closed loop gains down to unity, ×10 being the minimum recommended stable gain.

I mounted a sample of the device on the company's evaluation board 730037 Rev. A which is a nicely laid out 4cm square double-sided board with PTH (through-plated holes), designed to take PC mounting SMA sockets. It is laid out for leaded components throughout, but I used 10nF chip decoupling capacitors instead of leaded 0.1μF types (in parallel with tantalums of course), as these could be mounted closer to the device pins. For the rest, axial components were used, with R_f and R_g set at 470R and 51R respectively, Figure 3.1(b), setting a demanded gain of ×521/ 51 = +20.2dB and giving an insertion gain (board in to board out) of 6dB less than this. Figure 3.1(c) shows the test set-up used for subsequent measurements. Figure 3.2(a) is a double-exposure, the right-hand trace showing a −40dBm 100MHz test signal applied to a spectrum analyser direct, the corresponding zero hertz marker being one division in from the left-hand side. The second trace (offset one division to the left) shows the same signal applied via the amplifier on the evaluation board, (0Hz at extreme left-hand side of the display), indicating a gain of a shade over 14dB. Figure 3.2(b) shows another double exposure. The right-hand trace shows the output from the amplifier when the +5V and −5V supplies were turned off, the external attenuator having been set to 0dB instead of 30dB. Thus there is 38dB of isolation at 100MHz through the amplifier when

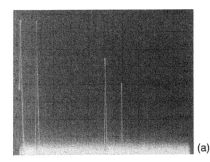 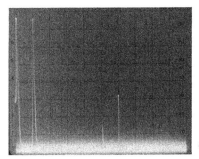

(a)   (b)

Figure 3.2 *(a) Right-hand trace, 100MHz −40dBm test signal applied to a spectrum analyser direct. (20MHz/div. horizontal, 0Hz one div. in from left-hand side, 10dB/div. vertical, ref. level 0dBm, IF bandwidth 300kHz, video filter off). Left-hand trace, as above but via the amplifier board, showing an insertion gain of just over 14dB. (b) Right-hand trace, as above but amplifier ±5V supplies turned off and the external attenuator set to 0dB instead of 30dB, showing 38dB of powered-down isolation. Left-hand trace, external attenuator set to 0dB instead of 30dB, amplifier powered up but input and output ports interchanged, showing 55dB reverse isolation and a 70dB ratio of forward to reverse gain.*

powered down. Even more interesting is the left-hand trace. Here the amplifier is powered up but has had its input and output ports interchanged. Again the external attenuator was set to 0dB instead of 30dB, indicating a reverse isolation at 100MHz of about 55dB and a ratio of forward to reverse gain of around 70dB.

With its low noise and flat gain versus frequency, an obvious application for the device is as a preamplifier to extend the input range of a spectrum analyser to lower levels. Spectrum analysers are designed to cope linearly with a welter of frequencies at their input, so as to display them all faithfully whilst adding the minimum of additional spurious 'signals' due to intermodulation products. They consequently usually employ a straight-into-the-first-mixer architecture, which whilst maximising linearity, results in a noise figure in the range 20 to 25dB. When a wanted signal applied to the analyser is known to be 'in the clear' without other large unwanted signals around, the sensitivity of the analyser can be extended by adding a low noise preamplifier at the input. To see the effect of the Figure 3.1 circuit in this application, a low deviation FM test signal was employed. The 100MHz test signal was frequency modulated with a 50kHz sinewave, with a modulation index of 0.14 – a peak frequency deviation of 8kHz and a peak phase deviation of only 8°. At this low modulation index, the second-order FM sidebands of the signal are almost 50dB down on the carrier, the amplitude of which is virtually unchanged from the unmodulated condition. Thus at not much over −110dBm, the second FM sidebands in Figure 3.3 (right-hand trace) are barely visible above the analyser's noise floor – which is itself almost half a division above the graticule baseline due to the smoothing of the 'grass' by the video filter. The insertion of the amplifier board between the external

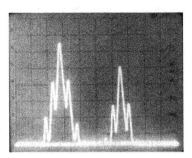

Figure 3.3   *Right-hand trace, at not much over −110dBm, the second-order FM sidebands of the 100MHz −60dBm test signal (see text) are barely visible above noise (100kHz/div. horizontal, 10dB/div. vertical, ref. level −30dBm, IF bandwidth 10kHz, video filter at max., 50kHz modulating frequency applied). Left-hand trace, via the 14dB gain of the CLC425 amplifier board, the second FM sidebands are clearly visible and their level readily measured.*

attenuator and the analyser's input, left-hand trace, rescues the signal and makes measurement easy.

Of course, the CLC425 is not the only solution in this application; a bespoke design using discrete transistors could doubtless achieve an even lower noise figure (though it must be said that the analyser's noise figure is so high that a second similar Figure 3.1 preamplifier stage could usefully be cascaded). But for an amplifier with 50Ω input and output impedances, high reverse isolation and a small-signal flat reponse from DC up to a few hundred megahertz, the circuit shown provides a no-design-time instant answer. Incidentally, on its ±5V supply rails, the device's maximum output swing for a THD of less than 1% is maintained at in excess of 7V pk-pk up to 10MHz, whilst the settling time to 0.1% is a shade over 20ns.

The other amplifier mentioned is the Texas Instruments TLE2027 (there is a premium TLE2027A version). The circuit schematic of this device, made on the company's Excalibur process, contains no less than 62 transistors, all bipolars except for one solitary FET. The TLE2027, internally compensated for gains down to unity, features the enormous open loop gain of 153dB (45V/μV) which, with its 15MHz unity gain frequency corresponds to an open loop gain roll-off starting at below 1Hz. (Contrast this with a gain roll-off in the other device starting at around 100kHz.) The TLE2027 offers excellent DC characteristics, the typical input offset voltage, tempco and drift being respectively 20 (10)μV, 0.4 (0.2)μV/°C and 0.006 (0.006)μV/month for the standard (premium A) version. As far as AC characteristics are concerned, the TLE2027 is clearly aimed more at the audio frequency range as against the video/RF applications catered for by the CLC425. The latter's equivalent input voltage noise (R_s = 50Ω) may actually be reduced by powering up slightly, to below 1nV/$\sqrt{Hz}$, against the TLE2027's 2.5nV/$\sqrt{Hz}$ (R_s = 100Ω), but with its 1/f corner frequency of 500Hz, the CLC425's noise has already risen to 8nV/$\sqrt{Hz}$ at 100Hz as against only 3.3nV/$\sqrt{Hz}$ at 10Hz for the TLE2027. As usual, it's horses for courses.

An unusual feature of the TLE2027 is its unusually wide phase margin, enabling it to tolerate large capacitive loads, Figure 3.4(a). Tolerate, yes, but not necessarily drive at anything other than a very low level. For the reactance of 10 000pF at 7MHz is a mere −j2.3Ω, and the device's internal current limit is of the order of ±35mA! In fact on ±15V supplies, the maximum output voltage of a little over 25V pk-pk into a load resistance R_L = 2kΩ is maintained over the full audio band and beyond, to 30kHz − Figure 3.4(b). With capacitive loading, the peak to peak swing available will clearly depend upon the load reactance, i.e. upon both the size of the capacitance and the frequency of operation. I was particularly interested in the device's ability to drive large capacitances, which turned out to be amazing, when a year or so ago I designed an RCL bridge. The circuit was based upon the transformer-ratio-arm principle (a scheme which has one

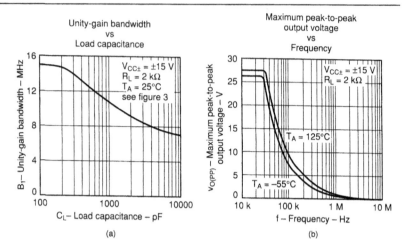

Figure 3.4   *(a) Unity gain bandwidth of the TLE2027 versus capacitive load. (b) The full power bandwidth of the TLE2027 covers the full audio band, and beyond. (Reproduced by courtesy of Texas Instruments Ltd.)*

unique advantage over any other parameter measurement technique) but realised without transformers by using op-amps instead. The arrangement required an op-amp which would drive capacitive loads of $1\mu F$, and even up to $10\mu F$ though at these larger values the voltage swing required would only be small. Most of the op-amps I looked at in the search for a suitable device would only cope with up to 100pF of load capacitance, often requiring even that amount to be buffered off from the op-amp output with a 47Ω resistor. But with Figure 3.4(a) showing the TLE2027 driving up to 10 000pF, this device seemed a hopeful candidate for the job – and so indeed it proved.

Figure 3.5(b) (upper trace) shows the TLE2027 driving 23V pk-pk directly into $1\mu F$ at 318Hz $(2 \times 10^3$ rad/s) in the circuit of Figure 3.5(a). The lower trace shows the residual THD (total harmonic distortion), which visibly contains both second and third harmonic components, and amounted to 0.07%. The frequency was raised to 500Hz, the amplitude being unchanged, at which frequency (Figure 3.5(c), upper trace) the impedance of the $1\mu F$ capacitor had fallen to 320 ohms. This resulted in the op-amp going into current limit over the negative-going flank of the sinewave – the data book showing the device current-sink limit being 35mA against a 40mA max. current source capability. The current limit does not come in at the peak voltage, as would be the case with a 320Ω resistive load, but over the part of the waveform where the slope dV/dT is greatest, due to the current drawn by a capacitor being $90°$ phase advanced. In the lower trace, the THD meter residual has been adjusted to show that the response is perfectly linear over that part of the cycle where

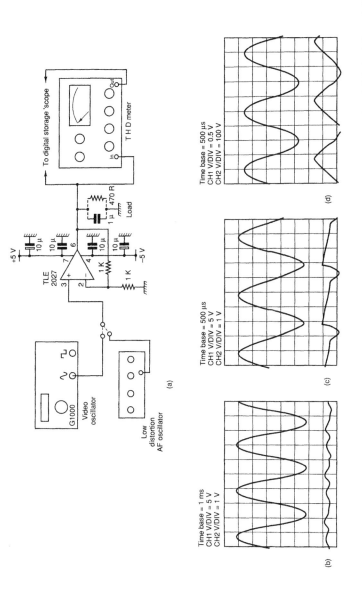

Figure 3.5 (a) Circuit used to test the TLE2027 driving capacitive loads. (b) Upper trace, TLE2027 driving 23V pk-pk into 1μF at 318Hz; lower trace, THD is 0.06% – but due to what? (c) Upper trace, TLE2027 (trying to) drive 23V pk-pk into 1μF at 500Hz; lower trace, showing the effect of the op-amp's internal current limiting circuitry. (d) Upper trace, TLE2027 driving 23V pk-pk into a 380 Ω resistor at 500Hz, lower trace, even at 0.005%, the 'residual' is still mainly fundamental.

the op-amp is not in current limit. This residual read 10% distortion, but clearly the waveform shown has a component at the fundamental, and when this was completely nulled out, the true distortion was measured as 6.3%.

This 500Hz measurement was repeated with the capacitance replaced with a 470Ω resistor, resulting in the op-amp not quite reaching current limit. The residual distortion is not shown as it was identical to that in Figure 3.5(b) not only in amplitude and waveform shape, but in *phasing*, a fact that started alarm bells ringing, since the phase of the load current was different. So the video generator supplying the drive (the same that was used to provide the 50kHz modulation in Figure 3.3) was replaced with an oscillator whose THD at 500Hz is less than 0.0005%. The voltage across the resistor is shown in Figure 3.5(d) (upper trace) and the 'residual' shown in the lower trace was 0.005%. In fact, it is largely fundamental, there being insufficient resolution on the THD meter's wirewound in-phase and quadrature trim potentiometers to achieve a complete null of the fundamental. This suggests that the distortion in the TLE2027 op-amp is not much if any more than the residual distortion of the distortion meter itself, which is known from other tests to be about 0.0016%. This is a remarkable performance for an op-amp supplying 23V pk-pk at a load current not much below the current limit value (given the 2kΩ gain-setting chain at the non-inverting input in parallel with the 470Ω load).

4 Current conveyor ICs – a new building block

There are unique benefits to current conveyor ICs in applications from LF to RF. A precision rectifier needing only two resistors and two diodes is just one example from many. This and various other applications are explored below.

Figure 4.1(a) shows a simplified diagram of the internal workings of what looks like, at first sight, an op-amp. True, an input to bias up the tail current has to be provided externally, but there is the usual NPN input

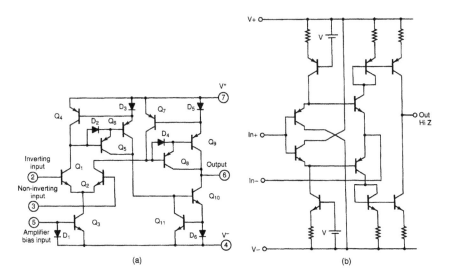

(a) (b)

Figure 4.1 *(a) Simplified circuit of an operational transconductance amplifier (OTA). (b) Simplified circuit of a current conveyor (CC).*

long tail pair, feeding into current mirrors hanging on the positive supply rail. With complementary emitter followers to provide a low output impedance unity gain buffer at the output, it would simply be an op-amp with facility for trading off supply current against speed. In fact, the buffer is missing, and the output is taken directly from two complementary transistors arranged as current sources fighting each other – the circuit is an OTA (operational transconductance amplifier) of the type that has been around for decades, e.g. the CA3080.

Figure 4.1(b) shows a close relative, another sort of transconductance amplifier but with a difference. Here, with a unity gain buffer at the output, you would have a current feedback op-amp (CFBO) – however, the buffer is missing so that the device has a very high impedance (current source) output just like an OTA. But instead of inverting and non-inverting (I and NI) inputs both of high impedance, this device known as a *current conveyor* or CC, has a high impedance NI input and a low impedance I input, just like a CFBO.

Just as an OTA (unlike an ordinary voltage feedback op-amp) is usually used without feedback, so is the current conveyor (unlike a CFBO). The absence of a feedback path avoids the stability problems that can plague voltage- or current-feedback op-amp designs, a welcome feature of the CC being its complete stability when driving reactive loads of either sign. But the device's current output requires a different approach to circuit applications, which cover the frequency range from DC up to 100MHz.

Figure 4.2(a) shows a simplified representation of a CC, the relation between the terminal currents and voltages, and the pin-out of the CCII01 device[1]. As with a CFBO, the CC's non-inverting (Y) input is high impedance (80kΩ at 1kHz) while the inverting (X) input is low impedance, see Figure 4.2(b). The CCII01, packaged as an 8-pin DIL, contains two current conveyors, and advantage can be taken of this to produce an enhanced composite conveyor with an input impedance at the X input up to 1MHz or so of less than 200 mΩ, Figure 4.2(c). With its low input impedance, the X node of either a standard or an enhanced CC can be used as a current summing junction for two or more signals, whilst its high Z-port output impedance (typically 1MΩ at a frequency of 1kHz) means that the signals from several CCs can be combined simply by hardwiring their outputs together. The notes contained in the data sheet give numerous other applications, including differentiators and integrators, voltage- and current-controlled NICs (negative impedance converters), precision half- and full-wave rectifiers, double-terminated amplifiers, etc.

One application is as a biquad filter and the data sheet shows the results obtained with the CCII01 configured as a 5MHz bandpass filter, shown also in Ref. 2. The equations defining the performance of the filter are discussed in the Box, and I experimented with the bandpass version at audio frequency, deliberately using extreme values of components to see if

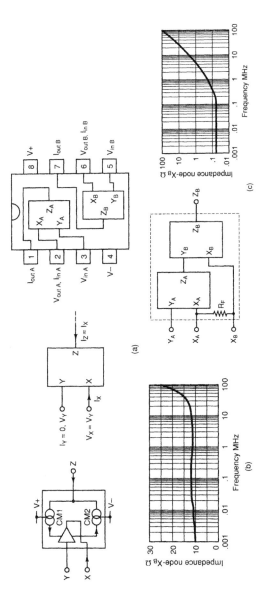

Figure 4.2 (a) Simplified representation of a CC, the relation between the terminal currents and voltages and the pin-out of the CCII01 device. (b) Input impedance of the X input as a function of frequency. (c) Input impedance at the X input of a composite current conveyor, as a function of frequency.

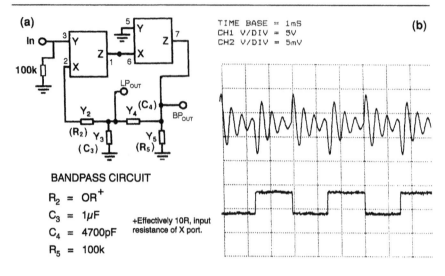

Figure 4.3 (a) Circuit of a bandpass amplifier, deliberately using extreme components values. (b) Output of the 2267Hz bandpass filter of (a), picking out the seventh harmonic of a 324Hz 4mV pk-pk squarewave.

the equations and performance still tied up. Figure 4.3(a) shows this extreme version, Y_2 being 0.1S, i.e. no actual R_2 was used, the 10Ω being the input impedance of the first CC's X input. From the equations in the BOX, the expected frequency peak was at 2250Hz while the measured peak occurred at 2267Hz. As the circuit used 20% capacitors, the close agreement must be taken as more a matter of luck than good engineering, and no such close agreement was found in the matter of Q. The theoretical Box value was 7.03 whereas the -3dB points at 2052Hz and 2495Hz indicate $Q = 5.1$.

The filter was then used to pick out the seventh harmonic of a 324Hz squarewave, the results being as in Figure 4.3(b). The Box value of the filter's centre frequency gain is 9950 or nearly 80dB, and this is reflected in the large difference in the Y deflection factors for the two traces. Given that the amplitude of the seventh harmonic of the 4mV pk-pk squarewave is about 0.73mV pk-pk, the average value of the filter output – at about 7V pk-pk – is what would be expected. The rate of decay of the damped wavetrain after the filter is excited by each edge of the squarewave (taking a time domain view for the moment rather than a frequency domain view) is described by its 'logarithmic decrement'. From this, using the relation 'Q equals energy stored over energy lost per radian' (an exceedingly useful result which is not as widely known as it deserves), one can calculate the Q. Careful measurements of successive peaks of the filter output waveform in Figure 4.3(b) show that the amplitude falls to 50% over each successive cycle. Thus the energy stored falls to 25% per cycle, or to 80% per radian. Hence the Q is (using an approximation only valid for high values of Q) about 100%/20% or $Q = 5$, agreeing with the measured value, mentioned earlier. Why the centre frequency and gain should agree with the theoretical values but the Q not, is something that quite escapes me for the moment.

In fact, Figure 4.3(a) is a very interesting circuit, for at DC (0Hz) the reactance of each capacitor is infinite, so that they effectively no longer appear in the circuit. Naturally, the gain at the low-pass output is then unity, since it is taken, via R_2, from the X port of the first CC, the output of the device's unity gain input buffer. At DC then, the circuit is completely open loop, so some output offset might have been expected at the bandpass output, given the high value of R_5 used. In fact, the output offset was zero, even with the 50Ω source disconnected leaving the Y input of the first CC grounded via 100kΩ. This says something for the accuracy of the device's fabrication, which is not carried out by LTP Electronics Ltd themselves, nor by a silicon foundry, but by one of the leading manufacturers of advanced linear ICs. This company may be offering the part under their own brand in due course, if it achieves sufficient design-ins to create a viable production volume.

With its 2000V/μs slew rate and 700MHz equivalent gain-bandwidth product, the device is billed as suitable for use at frequencies up to 100MHz. Now for any IC (or other active device) for use at high frequencies, an important parameter is its reverse isolation. For stable operation – especially when combined with tuned circuits – this should comfortably exceed the forward gain expected from any amplifier realisation using the device. The reverse isolation as a function of frequency was therefore measured, using the test set-up of Figure 4.4(a). The output of the DDS-based sweeper was 0dBm at low frequencies, falling linearly by no more than 1dB up to 100MHz. Figure 4.4(b) shows the reverse isolation over that frequency range, the spectrum analyser reference level being 0dBm and the IF bandwidth being set to a wide 3MHz in order to give a bright enough trace to register at each step of the sweep. Thus the reverse isolation can be seen to be in excess of 70dB below 15MHz, greater than 60dB below 70MHz and still 56dB at 100MHz. With this sort of performance, the CCII01 should enable stable, high gain RF amplifier stages to be readily realised.

This hypothesis was tested with the circuit of Figure 4.5(a). Here, a series tuned circuit at a CC's X port and a parallel tuned circuit (also at 10.7MHz) at its Z output is used to provide a two-pole response. Appropriate values of L/C ratio were selected for the two tuned circuits. The second CC in the package was used as shown as a high input impedance buffer to avoid loading the parallel tuned circuit. The −50dBm attenuated output of the sweeper was applied to the circuit, the output being as shown in Figure

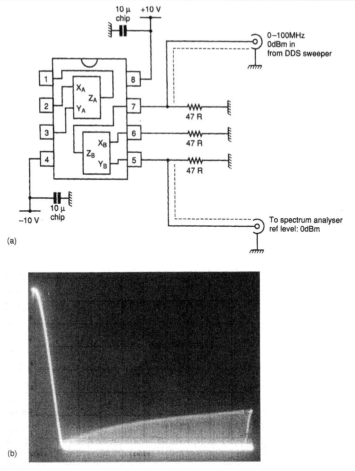

(a)

(b)

Figure 4.4   *(a) Measurement set-up used to test the reverse isolation of the CCII01 as a function of frequency. Sweeper output level 0dBm. (b) Results of the measurement over the range 0–100 MHz. Analyser reference level (top of screen) 0dBm, 10dB/div. vertical, IF bandwidth 3MHz, video filter off, 10MHz/div. horizontal, centre of screen 50MHz.*

4.5(b), indicating a gain of 30dB and a rather nice (if unintentional) bandpass response. Naturally, with a part having a response extending to the best part of a gigahertz, a compact layout using a groundplane was used, with the result that there was some coupling between the unshielded coils. However, the symmetry of the response indicates a lack of significant internal feedback in the stage, so the question that naturally arose was – can the circuit be pushed for even more gain and selectivity, without incurring instability?

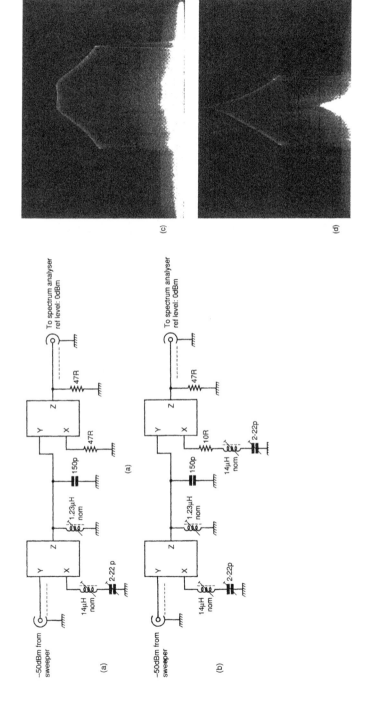

Figure 4.5 (a) Circuit of a 2-pole IF amplifier using the CCII01 current conveyor. (b) Circuit response, input −50dBm. Analyser reference level (top of screen) 0dBm, 10dB/div. vertical, IF bandwidth 10kHz, video filter off, 200kHz/div. horizontal, centre of screen 10.6MHz. (c) Circuit of a 3-pole IF amplifier using the CCII01 current conveyor. (d) Circuit response, input −50dBm. Analyser reference level 0dBm, 10dB/div. vertical, IF bandwidth 10kHz, video filter off, 200kHz/div. horizontal, centre of screen 10.7MHz.

The 47Ω resistor at the X input of the second CC in the package was replaced by another 10.7MHz series tuned circuit, all coils this time being of the screened variety. Occasional oscillation at around 400MHz was noted – yes, this really is a high frequency part – and was cured by inserting a 10Ω resistor between the X port pin 6 and the second series tuned circuit as shown in Figure 4.5(c). The resultant response was as in Figure 4.5(d), the screened coils having avoided any unintentional coupling. With the circuit now providing an output of -7dBm, gain is some 43dB, albeit as in 4.5(a) the circuit does not provide a 50Ω termination at its input. The slight asymmetry of the response, with a faster fall-off on the high frequency side, indicates a small degree of internal feedback. Perhaps not surprising, since although the reverse isolation of the CC at pins 5, 6 and 7 was measured as in excess of 70dB at 10MHz (I assume the other CC is similar), the two CCs are in very close proximity not only in the same package but also on the same chip. The remarkable thing about the circuit of Figure 4.5(c) is the almost total absence of discrete components, apart from the tuned circuits. Presumably additional tuned circuits, providing 50Ω interfaces to the chip, could be added at input and output, to give a 5-pole response.

Of course, one would probably never need in practice to design a 10.7MHz amplifier using individual tuned circuits as in Figure 4.5(c), since a wide range of block filters covering almost any conceivable requirement is available from numerous suppliers. But if an IF amplifier operating at a non-standard frequency is required, where ready-made filters are not available, then the current conveyor circuits shown provide a convenient way of providing the gain and necessary selectivity with the minimum component count.

Figure 4.6 shows just two other applications of the CC, not mentioned in Ref. 2. Figure 4.6(a) shows the dual of the voltage-controlled or short-circuit stable NIC shown in Ref. 2, namely a current-controlled (open-circuit stable) negative impedance converter. To see how it works, imagine Z is a 1kΩ resistor with its lower end grounded, the voltages at ports X, Y and Z are all initially zero, and keep in mind the relations between the port voltages and currents defined in Figure 4.2(a). Now, raise X to +1V. Y must also be at +1V since $V_x = V_y$, so there must be +1V at the top end of the 1kΩ resistor. So 1mA flows out of port Z and therefore 1mA also flows *out of* X, since $I_z = I_x$. This is the *opposite* of what would happen if port X looked like a 1kΩ resistor – it behaves just like −1kΩ. The lower end of Z was assumed grounded purely for purposes of explanation: in view of the device's high common mode rejection (greater than 53dB up to 1MHz), neither end of Z need be grounded, the circuit offering a floating negative impedance.

The very high output resistance of the Z port (typically 1MΩ at 1kHz) makes it behave like an almost perfect current source, forcing the demanded current through the load willy nilly, regardless of the volt drop across it. This is just what is required to deal with diodes in a precision

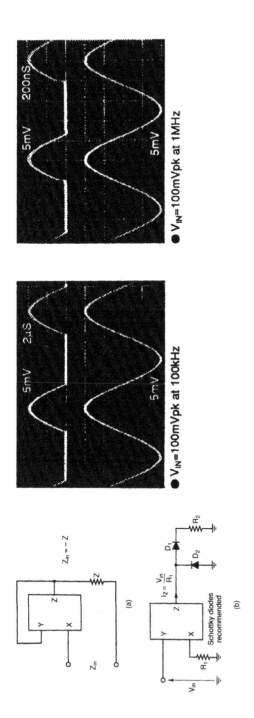

● V_{IN}=100mVpk at 100kHz

● V_{IN}=100mVpk at 1MHz

$Z_{in} = -Z$

(a)

$I_Z = \dfrac{V_{in}}{R_1}$

Schottky diodes
recommended

(b)

Figure 4.6 *(a) A current-controlled (open-circuit stable) negative impedance converter (see text). (b) A precision half-wave rectifier. (c) Performance of the precision half-wave rectifier in (b).*

rectifier circuit, Figure 4.6(b) showing a half-wave version. If $R_1 = R_2$, then the output is identical to the input for positive voltages and is zero for negative voltages, regardless of waveform or mark/space ratio. The results with a sinewave are shown in Figure 4.6(c), while the data sheet gives also a full-wave precision rectifier.

Readers wishing to experiment with the CCII01 should be wary of the simulated grounded inductance shown in the data sheet and in Ref. 2. Whilst this will indeed draw a lagging current from a zero resistance source, since both the CCs in the loop are non-inverting, the circuit looks like a *negative* resistor at DC, unlike a real inductor. In other words, the circuit is not open-circuit stable and will lock up at one or other supply rail.

Acknowledgements

Several of the illustrations are reproduced by courtesy of LTP Electronics Ltd.

References

1. The CCII01 is available from LTP Electronics Ltd, 2 Quarry Road, Headington, Oxford OX3 8NU. Tel. 01865 744232.
2. See 'Current conveyor circuits', *Electronics World and Wireless World*, Nov. 1993 pp. 962 and 963.

Box

Figure 4.3 shows a low-pass/bandpass filter using current conveyors; interchanging the resistors and capacitors gives a high-pass version. The transfer function of the circuit, in terms of the admittances of the passive components, is

$$\frac{V_{BPF}}{V_{IN}} = \frac{-Y_2 Y_3}{Y_5(Y_2 + Y_3 + Y_4) + Y_3 Y_4}$$

where Y_2 is the conductance $1/R_2$, Y_3 is the susceptance sC_3, etc., and s is the complex frequency variable. For the purposes of steady state analysis, s may be replaced by $j\omega$, and rewriting the equations in terms of C and R gives

$$\frac{V_{BPF}}{V_{IN}} = \frac{-sC_3 R_5}{s^2 R_2 R_5 C_3 C_4 + sR_2(C_3 + C_4) + 1}$$

Comparing this with the archetypal form for a bandpass filter

$$\frac{V_{BPF}}{V_{IN}} = \frac{As}{s^2 + Ds + 1}$$

the peak response will occur when the phase shift is zero (or 180° in this case, due to the minus sign in the numerator; the circuit is an inverting filter). This is so when the outer terms in the denominator add to zero, leaving just a $j\omega$ term to cancel top and bottom, i.e. a real (negative) number. Since $s^2 = (j\omega)^2 = -\omega^2$, this occurs when

$$\omega^2 R_2 R_5 C_3 C_4 = 1$$

Feeding in the component values from Figure 4.3(a) gives $\omega = 2\pi f_r = 14\,142$ whence the resonant frequency f_r is 2250Hz. At this frequency, the gain is

$$\frac{-C_3 R_s}{R_2(C_3 + C_4)} = -9950$$

and the Q is given by

$1/Q = D = sR_2(C_3 + C_4) = $ (in this instance) 0.142, whence $Q = 7.03$.

5 Single-pot polarity and gain adjustment

After this piece was published, I heard that the circuit appeared in a well-known weighty tome on electronics – a copy of which I do not possess and have not seen. Although not original to me, this useful little circuit is well worth passing on.

Using the 2 : 1 gain difference between inverting and non-inverting operational amplifiers, one potentiometer continuously varies this circuit's output from a replica of the input, through no output, to an inverted version, both at unity gain. It has many uses, for instance in audio processing. In conjunction with a BBD delay chip, it permits the position and amplitudes of the peaks or nulls of a comb filter to be adjusted.

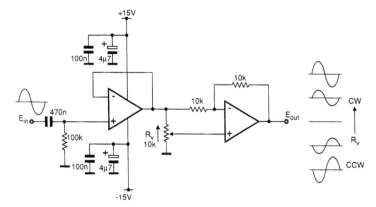

Figure 5.1 *Potentiometer R$_v$ varies output from inverted to non-inverted version of the input at unity gain.*

6 LCR measuring transformed

This article returns to the subject of the transformer-ratio-arm-bridge (see earlier article of that name), but with the source and detector interchanged – using the bridge 'back to front'. In this way, and discarding the transformers, an accurate LCR bridge with digital readout can be constructed.

A general-purpose component bridge covering a wide range of values of resistance, capacitance and inductance can be designed with just a few close tolerance resistors as standards, some op-amps and a little ingenuity. Furthermore, if the design is based, as the one presented here is, on the principle of the transformer-ratio-arm-bridge, then digital readout of the measured values is easily arranged. The transformer-ratio-arm-bridge was described in an earlier Design Brief,[1] where its use in the conventional manner was described, i.e. with the detector connected to the centre-tapped balance transformer. Being a passive linear network, however, it can also be used 'back to front', with the source connected to the centre-tapped balance transformer instead, Figure 6.1. In the simplified circuit

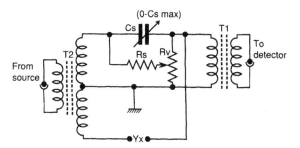

Figure 6.1 *Transformer-ratio-arm-bridge used with the usual source and detector connections interchanged. (In its simplest form, shown here, the TRAB measures only capacitance and conductance.)*

shown, only resistors and capacitors (lossy or otherwise) can be measured. Inductors can be measured either by arranging switching to connect C_s to the other end of the centre-tapped winding, or by connecting a fixed capacitor of value $C_{s.max}/2$ in parallel with the unknown susceptance Y_x: the shunt inductive component of Y_x is then measured as an equivalent negative capacitance. Similar arrangements for R_s enable negative conductance components of Y_x to be measured.

At balance, there is no current through the left-hand winding of T_1, and so no voltage across it. Thus this winding represents a virtual earth and T_1 could nowadays be replaced by the virtual earth at the input of a suitably fast inverting op-amp. It would be nice to be able to eliminate T_2 also: it turns out that not only is this possible, but one can actually eliminate C_s as well. Note that with the voltage applied to C_s in the phase shown in Figure 6.1, balance can be achieved with a capacitive Y_x, while with a voltage in the *opposite* phase applied to C_s (from the other end of the centre-tapped winding of T_2, the two halves of which are perfectly coupled), inductive unknowns are catered for. If a voltage in *quadrature* were available, it would be possible to balance a *resistive* Y_x against C_s, or alternatively (much more useful) a *reactive* unknown could be balanced using a *resistive* standard. It was this thought that gave me the basic idea for the design of a universal lab. LCR bridge, something I had been promising myself for some time.

To keep matters simple, fixed frequency operation at $\omega = 10^4$ rad/s (1.5915kHz) was chosen. That old favourite the SVF (state variable filter) was pressed into service as the oscillator, since it makes three outputs in quadrature available. As shown in Figure 6.2, it is simply a filter with zero damping, not an oscillator, but there is no practical difference between an

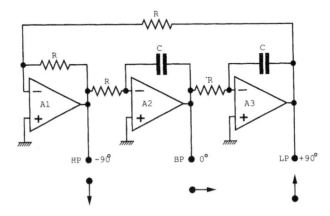

Figure 6.2 *A state variable filter with no input and no damping. Whether it oscillates or not is a moot point.*

oscillator and a filter with infinite Q. An integrator produces a 90° phase lag, e.g. the integral of a cosine wave is a sinewave, but each of the two integrators in the loop apparently produces a 90° LEAD. This is because they operate in the inverting connection, and the relative phases are thus as shown in Figure 6.2. The BP (bandpass) output has been labelled 0° because (at the filter's resonant frequency $\omega = 1/CR$) it would be in phase with an external input applied via a resistor at the inverting input of A_1. In the circuit shown, if R = 100K and C = 1nF, then the resonant frequency (frequency of oscillation) will be 1.5915kHz nominal. The circuit of Figure 6.2 was made up, using a TL084 quad op-amp, and it obligingly proceeded to oscillate at the required frequency – in fact at 1.5914kHz, which is surely good enough. With the ±12.5V rails used, the amplitude was ±11V peak, amplitude stabilisation being provided by slight clipping of the peaks in each op-amp. Evidently the excess phase shift in the op-amps (surely minimal at this frequency) together with layout strays ensured that the overall loop phase shift did not fall short of the 360° needed for oscillation, but on the other hand the very small degree of clipping indicated that it barely exceeded what was necessary. Had it not performed, a few picofarads in parallel with the 100K resistor from A3 output to A1 input would have persuaded it. Figure 6.3(a) shows the Lissajous figure, produced by the BP and LP (low-pass) outputs applied to an oscilloscope in X/Y mode, the clipping appearing as the slight flat tyres at the bottom and left-hand side of the circle. Both outputs measured 0.4% THD (total harmonic distortion). Although all three amplifiers were clipping, the BP and LP outputs each only show the clipping occurring in that particular stage. This is because the harmonics making up the dent in the input waveform to each integrator are attenuated much more than the fundamental at the output. However, the HP output from A_1 does not benefit

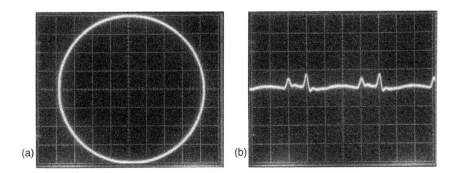

Figure 6.3 *(a) The BP and LP outputs of the oscillator of Figure 6.2 displayed as a Lissajous figure. (b) The THD meter residual when measuring the distortion of the HP output.*

from this, so both its own clipping and that fed back from A_3 can be seen in Figure 6.3(b), which shows the distortion meter residual output when measuring the HP output THD, which was 0.12%. Here, the time base speed has been adjusted to 157 µs per division, corresponding to 90°. That the clipping in A_1 and A_3 occurs in quadrature is clearly evident.

Having an almost faultless three-phase oscillator as the bridge source, work could now start on the pseudo-transformer-ratio-arm-bridge proper. The basic principle is simple, as illustrated in Figure 6.4. A 0° phase signal is applied to the standard resistance R_s via R_v and A_6 and thence to the virtual earth at the input of A_5. Thus the effective value of R_s is adjustable over the range infinity down to $R_{s.min} = R_s$, i.e. from zero conductance up to $G_{s.max}$. In the resistance position of S_1, a 180° signal is applied to the unknown terminals Y_x, from the unity gain inverting amplifier A_4. When R_x is a resistance equal to the effective value of R_s (a conductance Y_x equal to $1/G_s$ effective), the bridge is balanced, all of the current via R_s is just swallowed via Y_x by A_4, and no current flows in or out of the virtual earth of A_5. Therefore the detector registers no signal, indicating the point of balance. As R_s is effectively variable from infinity down to $R_{s.min} = R_s$,

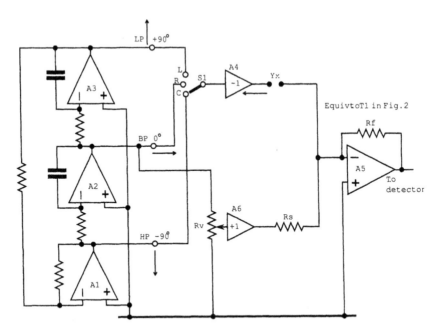

Figure 6.4 *This simple arrangement measures capacitance directly, but measures resistive and inductive impedances as the corresponding admittances. It also has no facilities for balancing out any reactive component when measuring a resistance or any loss term when measuring C or L.*

unknown resistors down to the same value can be measured. If an at-tenuator with steps of ×1, ×10, ×100 . . . is fitted between S_1 and A_4, resistors down to $R_{s.min}/10$, $R_{s.min}/100$, etc. can be measured, extending the range of the bridge to much lower values of resistance, whilst at the same time, keeping down the output current demanded from A_4.

If the output of A_4 is made to lag by 90° on that shown, by selecting the C position of S_1, then the (leading) current through the capacitor will again be in antiphase with that via R_s, enabling capacitive susceptances to be measured. Similarly, in the L position of S_1, inductors can be measured.

Whilst the arrangement of Figure 6.4 operates very like a transformer-ratio-arm-bridge, it has a number of drawbacks for use as a general-purpose component bridge. For example, it will work fine for pure resistors, capacitors and inductors: but there is no provision for nulling out the self-capacitance or self-inductance of a resistor, or the loss component of an inductor or capacitor. Also, although R_v can be calibrated directly in terms of conductance, giving a nice linear scale, a direct reading resistance scale would be more useful. Whilst a reciprocal scale could be used, reading resistance directly, it would be very open at low values and very very cramped at high values. On the other hand, Figure 6.4 will read capacitance values directly – as the capacitance at Y_x is increased, R_v must be advanced pro rata, not pro reciprocal, to maintain balance. Thus (apart from some provision for nulling the 'minor term', i.e. the quadrature or loss component) Figure 6.4 is basically what is required for capacitance measurements. For resistance and inductance, the variable facility, R_v and A_6, need moving to a position between S_1 and A_4, along with the ×10 step attenuator. Now, when R_v is set to zero, zero voltage is applied to Y_x, and so resistance and inductance will read directly on a linear scale fitted to R_v.

At the expense of slightly more complicated switching, all of the fore-going can be arranged, as shown in Figures 6.5 and 6.6. In Figure 6.5, note that an 11V pk-pk inverted (−180°) version of the 22V pk-pk 0° output of A_{1b} is always provided by the inverting amplifier A_3, and that a quadrature component, leading or lagging as selected by S_3, can be added to it as required. In the *resistance* position of S_1 this waveform is applied to the 1MΩ resistance standard STR (R_{15}), which is connected to the virtual earth at the input of A_5. Meanwhile, a 0° 11V pk-pk sinewave appears at the top of 10K potentiometer R_{13} and a proportion of this is applied via A_4, the ×10 step attenuator S_2 and A_2 to the unknown resistor at the test terminals Y_x. With S_2 in position 1, resistors from 1MΩ down to zero can be measured, with good resolution as R_v is a 10-turn potentiometer fitted with a digit dial. Nevertheless, the resolution would be too limited to measure accurately resistors in the low kilohms and ohms range, so other ranges can be selected by S_2, down to 0–10Ω max. in position 6. In the event that the 'resistor' under test has a significant reactive component, a deep null, indicating complete balance, will not be achievable with R_v

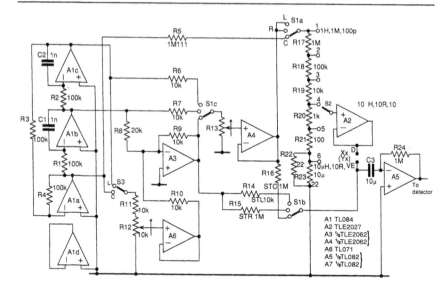

Figure 6.5   *Circuit diagram of the LCR component bridge and its 1.59kHz three-phase source.*

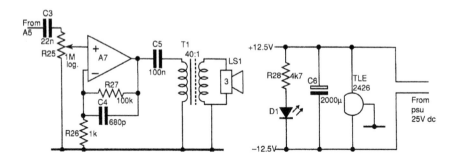

Figure 6.6   *Circuit diagram of the bridge detector and supply.*

alone. In this case, the minor or quadrature term can be nulled by adding a cancelling quadrature component, by advancing R_{12} CW (clockwise), i.e. in the direction indicated in Figure 6.5. As, at maximum, this quadrature component can equal the in-phase component at the output of A_3, the instrument can measure 'resistors' with a phase angle up to 45° – and, of course, the output from A_3 will then be 3dB greater than 11V pk-pk.

Inductors are measured in exactly the same way as resistors, with just two differences. To allow for the 90° phase lag of the current relative to the applied voltage, when measuring inductors S_{1c} selects the output of A_{1c}, which is advanced by 90° relative to the A_{1b} output which was used for resistive unknowns. As before, S_3 and R_{12} allow for phase angles up to 45° from the ideal, i.e. for an inductor Q of down to unity (or even an inductance with a shunt negative resistance component!). The other difference concerns the inductance standard STL (R_{14}). Figure 6.5 shows a value of 10kΩ, which provides inductance ranges of 0–1H down to 0–10µH.

Capacitance measurements are made rather differently. Whereas for both R and L, the voltage applied to the unknown was adjustable both in steps (S_2) and continuously (R_v) with a fixed voltage applied to the standard, for C measurements the variable voltage is applied to the standard STC (R_{16}) while the voltage applied to the unknown capacitor is varied only in ×10 steps. And to allow for the leading nature of the current through a capacitor, S_{1a} selects the lagging voltage from A_{1a} in place of an in-phase voltage. The result of this rearrangement is that again the digital read-out dial of R_v reads the value of the unknown C directly, as it did for R and L. R_{12} provides for balancing the capacitor's loss component, down to a tan δ of unity, or of course for a capacitive susceptance including a negative conductance component. A_2 must be a special breed of op-amp; it must be capable of driving capacitances up to 10µF. Many op-amps get very unhappy when faced with large capacitive loads – 'large' in this context may mean a few hundreds or even a few tens of picofarads – and may actually oscillate unless special precautions are taken. Here, however, there are no problems, as A_2 is that remarkable op-amp the TLE2027. This was described in an earlier Design Brief,[2] where it was shown driving 23V pk-pk into 1µF at 318Hz. Here, it is required to drive up to 10µF at 1591Hz, but only at 110µV pk-pk, in position 6 of S_1, or up to 11V pk-pk in position 1 where the maximum capacitance load is only 100pF.

Any unbalance of the bridge results in current flowing in the virtual earth of amplifier A_5, which thus provides a signal to the detector stage, shown in Figure 6.6. A 1MΩ log. pot. precedes an amplifier A_7 (with 40dB gain), which drives a loudspeaker. A 3Ω loudspeaker with output transformer was pressed into service as it was to hand, but a reasonably sensitive 64Ω loudspeaker would do as well. The 2000µF capacitor C_6 provides additional smoothing for the 25V DC power supply, which is split into ±12.5V supplies for the op-amps by the TLE2426 'rail splitter', an ingenious Texas Instruments device that I have used in a number of projects. An LED 'ON' indicator was fitted, in the form of D_1. With the bridge measuring unknowns of both high and low impedance, it was desirable to keep both electrostatic and magnetic hum fields out of the instrument's metal case. One of the very inexpensive DC supplies built into a 13A plugtop style case was therefore used. As supplied, the circuit was as in Figure 6.7(a), but this was modified by removing the output voltage switch, which made room (just) for an additional 470µF capacitor. After modification, the supply was as in Figure 6.7(b), its output permanently connected

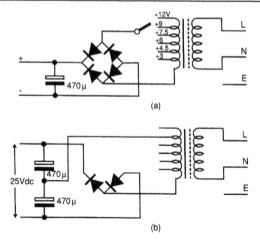

Figure 6.7   (a) Circuit diagram of a commercial 'plugtop power supply' as bought. (b) Circuit diagram of the commercial 'plugtop power supply' as modified.

to the case of the bridge by a length of common or garden audio screened lead.

This pseudo-ratio-arm-bridge proves delightfully simple to use, providing a resolution of 0.1% of full scale on any range, thanks to the 10-turn digit dial on R_v. When measuring resistors, the setting of the quadrature control R_{12} will be at or very near zero. The null obtained is deep and complete, with no sound audible in the loudspeaker other than a very slight trace of mains hum. This is due to the absence of stabilised supplies, which are – as it turns out – quite superfluous. When measuring capacitors or inductors, just as deep and complete a null is obtained as far as the fundamental is concerned, once the loss component has been nulled out with R_{12}, S_3 being in the C or L position as appropriate. However, as the reactance of a capacitor at harmonics of the drive waveform differs from that at the fundamental, whereas that of the capacitive standard STC (R_{16}) does not, some audio output at harmonic frequencies remains at balance. The ear easily distinguishes between the fundamental and the harmonic tone and thus a true balance is readily achieved. When measuring an inductor, as its reactance *rises* with frequency, harmonics are not accentuated and (as with resistors) are inaudible at balance. Setting STL (R_{14}) at 10kΩ (rather than 1MΩ as for STR and STC) limits the maximum inductance that can be measured to 1H, but provides ranges down to 10μH maximum, permitting in principle measurements down in the nanohenry range. To test low inductance measurements, four turns were wound on a two-hole balun core type FX2754 (which uses 3C85 material). The expected inductance, given the core's A_l of around 3500nH/

turn, was 56µH. The measured value, on range 5 of S_2, was 52µH with the dissipation or quadrature control R_{12} set near to zero, indicating a high value of Q. The number of turns was then reduced to one, with an expected inductance of 3.5µH. On range 6, the measured value was 3.9µH, with R_{12} set at about 30%, indicating a Q of about 3. Ideally, the measured value would have been one sixteenth of 52µH or 3.24µH, but the measured value will of course include the inductor's leads and the bridge's terminals and internal test circuit wiring. Remember also that this bridge, like the transformer-ratio-arm-bridge from which it is derived, measures an unknown as a *parallel* combination of susceptance and conductance. When the Q is low, the inductor is effectively an inductance and a resistance in *series*, i.e. the impedance is higher than the reactance of the inductor alone. Measured in *parallel* terms, this will appear as a rather *higher* value of inductance, in parallel with an even higher value of resistance. If the quadrature control R_{12} were calibrated, a series/parallel conversion could be applied to the results, to obtain the actual value of inductance, and its effective series resistance – and thence its Q.

This does, however, point up a limitation of inductance measurements carried out at such a low frequency as 1.59kHz, common to all bridges operating at $\omega = 10^4$. In fact at this frequency the Q of an air-cored inductance (or one with a slug but a return path in air, such as an RF choke) will be so low – less than unity – that balance will not be obtained in the L position of S_1, but only in the R position. On the other hand, small mains transformers may have a primary inductance of many tens of henrys. This being so, one might find it more convenient to use the 1MΩ STR also as the inductance standard STL, giving inductance ranges from 0–1mH up to 0–100H. An even better scheme would be two 'L' positions on S_1, providing inductance ranges all the way from 0–10µH to 0–100H. Extending the idea even further, an alternative value of 10kΩ for STC would extend capacitance measurements up to 1000µF.

The excellent performance provided by the bridge was not obtained without a little attention to construction. The recommended precautions are few, but necessary. Firstly, all the earth returns shown in Figure 6.5 (other than those associated with A_1) must be routed to a single star earth point, which can conveniently be the N (neutral) terminal; this was situated on the front panel, between and slightly below the D and VE (drive and virtual earth) terminals. Secondly, resistors associated with S_1 and S_2 should be mounted on the switches themselves and, untidy though it may look, connections from the switches and R_{12}, R_{13} routed directly in fresh air to the appropriate points on the circuit board – definitely no neat cableforms. Thirdly, the board layout should be such that A_2 output is as close as possible to the rear spill of the D terminal, say less than an inch of stout wire, and S_{1b} wiper should be returned direct to the rear of the VE terminal, from whence the lead to C_2. As far as accuracy is concerned, all resistors should be 1% or better, but more importantly R_6 and R_7 should both equal R_{13}, R_{11} should equal R_{12}, R_9 and R_{10} should each be half of R_8, and R_5 should equal the resistance from S_{1a} wiper to ground. The SVF oscillator frequency should of course be as close to $\omega = 10^4$ as possible.

The bridge was destined for use as a lab. general-purpose component bridge. But nevertheless, designed and constructed as described, the instrument should share the same unique attribute as the transformer-ratio-arm-bridge, of measuring without error the series component of a pi network whose shunt arms are grounded. This was verified as follows. A 56pF capacitor was measured, the value reading 56.5pF. 100pF capacitors were then connected from D and VE to N and the measurement repeated. The result was 56.4pF. So much for high impedance circuits; the test was repeated as follows. An 8R2 resistor was measured, the reading being 8R35. 4R7 resistors were then connected from D and VE to ground and the test repeated. The sensitivity was noticeably reduced, but interpolating between the points at which the tone just reappeared each side of the null gave a reading of 8R30. This verifies that for both high and low impedance circuits, the instrument can measure the series element of a pi network, even when the shunt arms present a lower impedance than that being measured. Thus, for example, a component on a PCB can be accurately measured without disconnecting it, if the far ends of other components connected to it are grounded.

As a lab. instrument, getting occasional use, the bridge has proved very satisfactory. For more concentrated use, especially by unqualified personnel, an automated version would be preferable. Such a scheme could be readily implemented using VCAs (voltage controlled amplifiers), as follows. The detector stage R_{25} and A_7 would be replaced by two synchronous detectors, one driven from the 0° phase and one from the 90° phase. The DC outputs of the synchronous detectors would be filtered and amplified, and fed back to control two VCAs, fed with the said 0° and 90° signals. Given that the VCA output was linearly proportional to the control voltage, as is the case for a four-quadrant multiplier, the two control voltages represent the real and imaginary components of the unknown directly, and could be indicated on DPMs (digital panel meters). It is true that the results will be the components of the equivalent shunt representation of the unknown, but often this will not matter. For resistive components with a phase angle at the test frequency of less than 5.3°, capacitors with a tan δ of less than 0.1 and inductors with a Q greater than 10, the error in the value of the component as indicated will be less than 1%.

References

1. Hickman, I. 'The transformer ratio arm bridge', *Electronics World and Wireless World*, August 1994 pp. 670–672.
2. Hickman, I. 'Differentiating op-amps', *Electronics World and Wireless World*, April 1994 pp. 322–325.

7 Cautionary tales for circuit designers

Some circuits look plausible – but in fact don't work, some work in spite of appearances and some could work with a following wind. In some cases the snag, if there is one, is obvious; in others it is much more subtle. A selection of such circuits is described below.

A glance at a circuit diagram is often enough to tell one what it is supposed to do, and just a little further thought will usually enable one to judge whether it will actually do it. Sometimes, though, there is a hidden catch, and the circuit won't work: other times, it turns out that a seemingly unlikely circuit *will* work. Here is a selection of circuits for you to ponder, most but not all falling into the former category, which I have collected over the years.

The first is a scheme for deriving an equal mark/space ratio squarewave from one with an unequal ratio. Everyone knows that if you divide a frequency by 2 you get an equal mark/space ratio, right? But Figure 7.1(a) is an incredible howler that anyone can see through almost immediately. Well, almost anyone, as it was submitted by someone who presumably thought it would work (although he obviously hadn't tried it) to the readers' design ideas section of one of the controlled circulation magazines (now defunct). The magazine's editor also presumably thought it would work, as it duly appeared in print: I carefully checked the magazine's date – it was *not* the April issue! Many readers wrote in to say it doesn't work, one submitting two alternative circuits that do, also shown in Figure 7.1. The circuit of Figure 7.1(b) operates over at least a 10:1 frequency range (given the appropriate component values), delivering a 50/50 ratio output. That of Figure 7.1(c) also operates over a range of 10:1 or more, with the further advantage that one edge of the output squarewave is coincident with that of the asymmetrical input waveform.

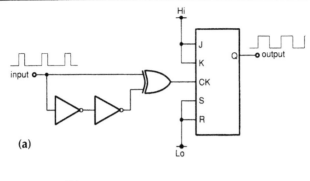

(a)

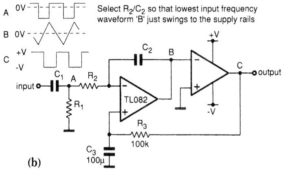

(b)

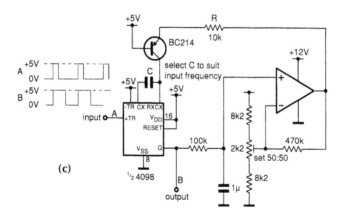

(c)

Figure 7.1 *(a) Proposed circuit for deriving an equal mark/space ratio squarewave from an asymmetrical one. (b) Alternative circuit that works. (c) Another alternative, this one having an edge that is coincident with the input waveform.*

Next, a circuit from the dim and distant past – the early 1960s – before the introduction of even SSI, when logic circuitry still used discrete components. I and my colleagues were working on a missile test equipment which was the first to make extensive use of digital measurement techniques, all measurement results, whether volts DC, volts AC rms, frequency, period, dV/dt or whatever, being read out on a purpose-designed DVM, also part of the project.

A colleague charged with designing part of the digital circuitry had an arrangement of gates which was something like that shown in Figure 7.2(a) using various diode logic gates of the type illustrated in Figure 7.2(b). It incorporated the bright idea of feeding back a gate output to an earlier gate, to which was applied a short 'take measurement' command pulse. This neatly ensured that the output gate was held open for the duration of the measurement, however long that took, depending upon the particular measurement type. Unfortunately, it didn't; even substituting a transistor AND gate at gate B didn't help and a discussion ensued amongst us all as to why not. I pointed out that the 'gain' through a diode gate or even an emitter follower gate was just a little less than unity, so that after passing through a couple of gates, when the signal was fed back to an earlier one, it was impossible for it to hold itself on. Our colleague went away to think about it and decided that the answer was to include an inverting transistor gate as shown in Figure 7.2(c). These were only used where essential on cost grounds, but here it would serve to include the necessary gain in the loop and so was justified. Unfortunately, it had the incidental property of inverting the logic signal, so that didn't work either.

Finally he came up with the solution: the input logic was allowed to be an inverting transistor gate (requiring an inverted 'end of measurement' signal) whilst gate B would be similar. It worked a treat. 'Congratulations', said someone, 'you've just re-invented the Eccles Jordan flip-flop!'

Now for another lame-duck circuit; one which was actually proposed (in an article about bootstrapping, by someone who should have known better) in the august pages of this very magazine, quite a few years ago when op-amps were less common and discrete transistor circuitry still the norm. The scheme for bootstrapping the base bias circuit, Figure 7.3(a) is well known and very effective, especially if the load on the emitter follower's output is light and it is provided with a constant-current long tail so that its gain is very very close to unity. The input resistance will still include a shunt contribution from the transistor's collector/base resistance, but if the collector voltage were to follow the emitter voltage (and hence the input signal voltage), this component would be bootstrapped out of sight also. What could be simpler, as shown in Figure 7.3(b).

Unfortunately, it obviously cannot work, for if the base current is negligible and the load on the circuit's output likewise, then the collector current must at every instant equal the emitter current. So when the drop

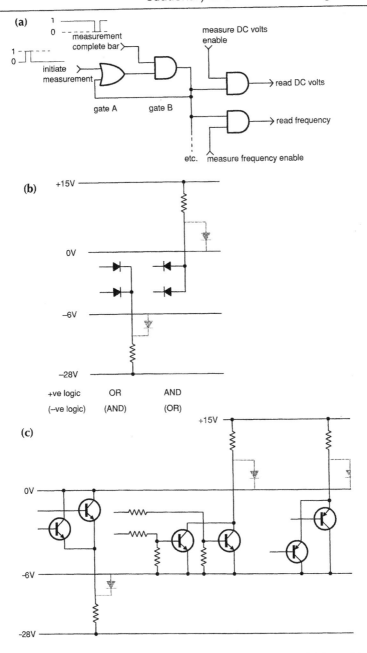

Figure 7.2 *(a) This logic gate arrangement in part of an early DVM would work with any modern logic family but not with: (b) diode logic gates have a voltage gain of not quite unity; (c) transistor gates; this 2 input OR gate also has a gain of just less than unity but the NOR gate has internal gain.*

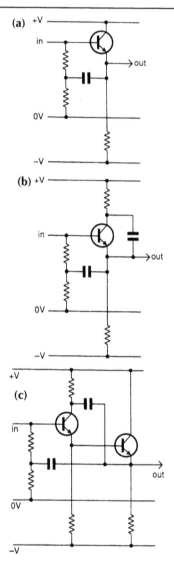

Figure 7.3   *(a) Bootstrapping an emitter follower's bias circuit to raise its input impedance works. (b) Bootstrapping its own collector for the same purpose doesn't, unless: (c) there is additional current gain in the loop, as here.*

across the emitter resistor increases, so must the drop across the collector resistance; there is nowhere else for the extra emitter current to go. Ergo, the collector voltage must fall. What if the emitter follower has a constant-current tail? In this case, the collector current cannot change and so neither can the voltage drop across the collector resistor, but the current through the transistor can change. The emitter can only follow a positive-going input by discharging the collector bootstrap capacitor. On a nega-tive-going edge, the transistor will cut off and the constant tail current transfer to the capacitor, charging its bottom plate negative-wards until the transistor cuts on again, whilst the collector voltage remains undisturbed.

Collector bootstrapping to get rid of the reduction of input impedance due to r_{cb} is, however, useful and effective, it just requires some extra current gain, to allow the input emitter follower's collector current to differ from its emitter current as in Figure 7.3(c).

This point about the equality of an emitter follower's emitter and collector currents is often overlooked. At one time I was required to design a calibrated variable phase shift circuit as part of the test facilities in a (later model of) missile test equipment. Having been impressed by an article called 'The Selectoject' – a circuit providing a tunable audio frequency bandpass or notch characteristic, as required – in the (then) *Wireless World*, I borrowed the basic idea and used the circuit shown in Figure 7.4. Since the emitter and collector resistors are equal, the voltages at those electrodes must be equal in amplitude and in antiphase. When the reactance of C numerically equals R, the current in the branch RC will be leading the voltage across it by 45°. Thus the output voltage will be in quadrature with the input and variable over the range −180° through −90° to 0° degrees as

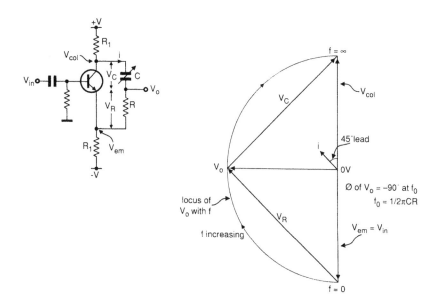

Figure 7.4 *Simple all-pass filter stage with unity gain at all frequencies and a phase-shift varying from 0 to −180° as C varies from zero to infinity, or as the frequency varies from zero to infinity for a fixed value of C.*

C is varied from infinity down to zero. The range actually available was of course less than that, but the circuit worked very well (considering the performance of the transistors then available) given that the operating frequency was actually well above the audio range. But naturally its performance wasn't quite perfect. Said a colleague, 'I don't see how you can expect it to be, given the unbalanced source impedances driving the ends of the series CR. The bottom end is driven by the low output impedance of an emitter follower and the top end by the collector output impedance, which is high.' So here is a circuit that *will* work, even though at first glance one might think it wouldn't.

Some circuits work at times but not at others, that is to say one example works but another build of the same design does not. Figures 7.5(a) and (b) look like a 'spot the difference' puzzle, the only difference being the addition in (b) of D_3. Figure 7.5(a), which appeared in the readers' design ideas section of one of the controlled circulation magazines, is a stabiliser circuit designed for use with a bank of NICAD cells. Although these have a fairly constant voltage during the discharge cycle, there is some voltage sag, especially if the operating temperature range is wide. This is undesirable if the battery pack is powering sensitive measuring equipment, so the stabiliser circuit shown was developed.

The compound pass transistor stage Tr_3/Tr_4 is controlled by the error amplifier Tr_1, which compares the fraction of the output voltage across $R_4 + R_5$ (part) with the reference voltage across zener diode D_1. Regulation is good, and so is stabilisation, since the reference voltage across D_1 (which should be a type with a sharp knee, suitable for use at low current) is derived from the stabilised output rather than the raw supply. The consequence of this, however, is that the circuit is bistable: if no output voltage then Tr_1 is off, and so no drive to Tr_3 and if no drive to Tr_3 then no output voltage. So R_2 is included to ensure start-up when the batteries are first connected, or following the removal of an extended short-circuit at the output. Tr_2 provides short-circuit protection by limiting the drive to the pass transistor when the output current reaches a level sufficient to drop about 600mV across R_1.

D_2 provides a path for recharging the bank of NICADs, for it was envisaged that the circuit could be incorporated actually within the NICAD battery pack. This prevents the danger of damage or even fire if the battery were accidentally short-circuited, since with large cells there is a lot of stored energy, and on short-circuit this can be released in a very short time. Housekeeping current on no-load is a miserly 55μA, but when the battery pack is not in use this can be reduced even further to a negligible 4μA or so (via R_2) if the battery pack be stored with the terminals short-circuited.

The clever part of the circuit is C_1, which isn't there just for show. As explained, on *extended* short circuit, the drain on the battery falls to a few

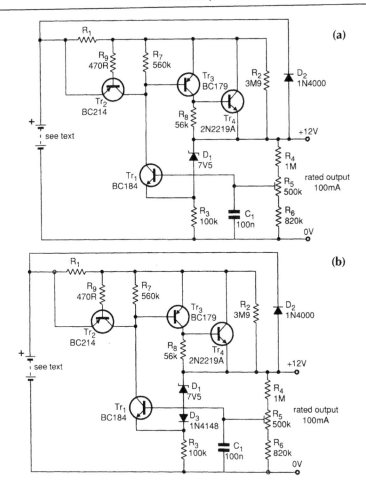

Figure 7.5 *'Spot the difference'. (a) Circuit of a current limited power supply whose output voltage may fail to recover when connected to a load including a large electrolytic. (b) The addition of D_3 cures the problem.*

microamps. But when the battery pack is connected to an instrument, it may find itself suddenly in parallel with a large decoupling electrolytic. This will cause the output terminal voltage initially to drop to zero, after which the capacitor will be rapidly charged up at the short-circuit current determined by R_1, *if and only if* Tr_1's collector is still supplying base current for Tr_3. C_1 fulfils this condition by maintaining the voltage at Tr_1's base long enough for the terminal voltage to recover to a level (a volt or two) from which it would then build up to the rated output anyway. The prototype circuits reliably turned on into a load including a 2000μF capacitor, so all seemed well. Some years later I had occasion to use this circuit again and built it up exactly as in Figure 7.5(a), only to find that on connecting a capacitor greater than a few tens of microfarads at most, the output voltage would not recover: the circuit remained sullenly switched off.

Solving this teaser took several cups of coffee before the light dawned. The stated purpose of C_1 is to hold up the voltage at the base of Tr1 while the short-circuit current set by R_1 (*not* the short-circuit current via R_2) started to charge up any external capacitance which might be connected. But unfortunately there is a discharge path for C_1, via the base emitter junction of Tr_1 in series with D_1 (working as a normal diode in forward conduction), and on via the momentary short across the output terminals due to the external electrolytic, back to C_1's other terminal. This was blocked by the addition of D_3, as in Figure 7.5(b), clearing the fault entirely. What was not clear, nor still is, is why the first prototypes worked, with as much external capacitance as one cared to throw at them, whilst the later ones would only stand a few tens of microfarads. Clearly a case of minor differences between characteristics in devices which are nonetheless all individually within specification. Would the problem have shown up on a CAD simulation package – Touchstone or SPICE, for example? I wonder. It depends on what limit values are built into the library models for the various parameters of the devices used, such as the extrinsic base resistance r_{bb}' of Tr_1, etc.

Now for a real peach of a circuit, Figure 7.6(a): if it really worked it would be extremely useful. By way of introduction, remember that if you want to make a high Q filter, be it low-pass, bandpass or high pass, you need at least two poles. And if we are talking about an RC active filter and want it tunable, that usually means a two-gang potentiometer or two-gang variable capacitor. Thus the CR product of the frequency determining sections can be varied in step, providing – say – a 10 : 1 tuning range for a 10:1 variation of the variable elements.

This is not to say that you can't make a variable frequency filter or sinewave oscillator using a single variable element; on the contrary you certainly can and Ref. 1 gives an example, while Ref. 2 describes no less than five such circuits. But the price you pay includes amongst other things, a reduced tuning range; an n : 1 variation of the tuning element R gives only a $\sqrt{n}:1$ tuning range. The Figure 7.6 circuit seems to break through this limitation. And it should work – given ideal components.

To see how it is supposed to work, it is best to take it in stages. Firstly, yes, the circuit is DC stable. There is feedback from A_1's output to its NI (non-inverting) input, but it is via A_2 which is inverting at DC so this loop provides negative feedback and is stable, whilst the loop through A_3 is DC blocked. Secondly, imagine the integrator and differentiator removed and A_1's NI input grounded. Then the circuit is inverting with unity gain, whatever the setting of R_Q. As the wiper of R_Q is moved toward ground,

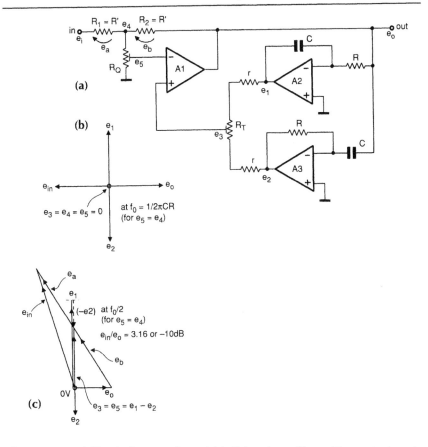

Figure 7.6 *(a) Circuit diagram of a variable Q bandpass filter with constant centre frequency gain of unity. (b) Vector diagram showing operation at the centre frequency, at minimum Q. (c) Vector diagram showing operation at one octave below the centre frequency, at minimum Q.*

the error voltage at the junction of R_1 and R_2 is attenuated more and more, but if A_1 is ideal, there will always be enough loop gain to ensure a gain of $\times(-1)$.

Now consider the case where the wiper of R_Q is at the top ($e_4 = e_5$), the wiper of RT is at mid-travel, and the input signal is a sinewave of frequency of $f_o = 1/(2\pi CR)$. At this frequency both the integrator and the differentiator have a gain of unity, the integrator output e_1 *leading* e_o by 90° (it is an inverting integrator) and the differentiator output e_2 lagging by 90°. Thus e_3, the net voltage at the wiper of the tuning control R_T, is zero and $e_o = -e_i$. If the wiper of R_T is moved towards the integrator output, e_3 will be zero at a somewhat higher frequency, or at a lower frequency if moved towards the differentiator output. In the limit, at the end of R_T's travel, for zero output at the wiper, one output must be $(R_T + r)/r = N$ times the other, i.e. at $\sqrt{N}$ times f_o. Thus the tuning range is from $f_o \sqrt{N}$ to $f_o/\sqrt{N}$, or $N:1$. As r is made smaller relative to R_T, the tuning range becomes larger and larger.

Figure 7.6(b) shows the situation at the bandpass centre frequency and Figure 7.6(c) at a frequency one octave lower, for the case where $e_5 = e_4$ (minimum Q). Clearly, as e_3 increases off-tune, so e_o becomes smaller relative to e_i, so e_4 is no longer zero. Still, off-tune the output will not be far below unity as long as e_3 is small compared to e_5. The allowable detuning whilst still meeting this condition gets smaller and smaller as e_5 becomes a smaller proportion of e_4. Finally, as the wiper of R_Q approaches ground and e_5 tends to zero, any departure whatever from exact equality of e_1 and e_2 (i.e. any departure from the exactly on-tune condition) will result in a fall in e_o. Put another way, in these circumstance, A_1 will produce whatever output is necessary to keep the signal at its NI input equal to that at its I input. If e_1 does not equal e_2, the only way it can arrange this is if e_o is near zero. The circuit provides a range of Q variable up to infinity, but with the on-tune response remaining at unity independent of the value of Q.

In principle all is fine; in practice the circuit is likely to oscillate – it certainly did when I tried it. The problem is the loop from A_1 to A_3 and back to A_1. Integrators are splendid, docile circuits, since the demanded (closed loop) gain falls with frequency at 6dB/octave, the same rate as the open loop gain of an internally compensated op-amp. Thus the gain within the loop is constant until way beyond the unity gain frequency and stability therefore assured.

Differentiators are a very different kettle of fish: the demanded gain rises at 6dB/octave, while the open loop gain falls at the same rate. Eventually the demanded gain exceeds the open loop gain and all bets are off. A_1's output is then effectively connected directly to A_3's input, with both op-amps contributing 90° of phase shift. At a high enough frequency, additional poles appear in the op-amps' open loop responses and oscillation results. Perhaps with a very high performance A_3 with a little capacitance across its feedback resistor and a little resistance in series with its input capacitor, one could turn it into an integrator at some frequency way above the band of interest, ensuring the stability of the circuit as a whole – an intriguing thought.

Finally, another very useful circuit. I have had it on file for some time but have not made it up myself. However, at a recent gathering of engineers I fell into conversation with someone who had, and he claimed it didn't work. Did he substitute different components or values, or just get the wiring wrong? Or is there really a problem? I can't see any reason why it shouldn't work in principle (though a tolerance exercise on the component values might not come amiss) so in my book it remains a definite maybe. The circuit, Figure 7.7, is an electronic mains fuse, but faster than a fuse, a

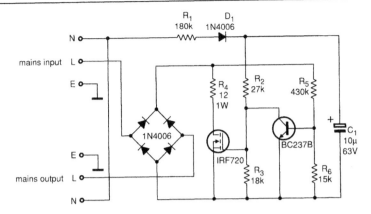

Figure 7.7 *Circuit of a fast-acting limiter allowing a maximum output of 100W, beyond which it exhibits a re-entrant foldback characteristic.*

thermal trip or a magnetic breaker. In fact, it is not so much a fuse as a limiter, since if the load tries to draw more than the rated 100W, the circuit exhibits a re-entrant foldback characteristic. Circuit operation should be clear from Figure 7.7 Such a device is clearly a must for the lab bench, so at the first opportunity I shall try it out. In the meantime, the evaluation of the viability of this circuit is (as it says in so many text books) 'left as an exercise for the reader'. (Hint: what voltage will the peak current through, say, a 125W resistive load drop across R_4 plus the power MOSFET, and is this enough to turn on the NPN transistor?)

References

1. Dean, A. P. 'Easily tuned bandpass filter', *New Electronics*, 19 Feb. 1985 page 24.
2. Williams, P. 'RC oscillators: single-element frequency control', *Wireless World*, Dec. 1980 pp. 82–84.

8 Analog signal processing

This substantial article appeared in two instalments, as 'Give ASP a chance', August 1995 pages 691–695, and 'Asp and filtering', September 1995 pages 787–791. It includes a neat scheme for extracting a repetitive or one-off signal from a source of repetitive noise (e.g. mains related) of a different frequency, a form of autocorrelation applied not to the wanted signal, but to the noise. It is reproduced here as originally submitted, as a single article with the illustrations numbered Figures 8.1 to 8.13.

When electrical signals (perhaps from a transducer of some kind) need processing to enhance the wanted information they contain, or to suppress unwanted interference, many designers will automatically turn to DSP. And indeed, digital signal processing can be as powerful as one wants (or as powerful as one's skill in writing algorithms permits), at a price. But very often, any necessary signal processing can be accomplished by analog circuitry – by ASP. The expense of writing algorithms, of ADCs and DACs, DSP chips and memory are thus avoided and a solution achieved with a much lower power budget, an important consideration in portable battery-operated equipments. The absence of clock signals can also be a boon, especially where small signals are being handled in a miniaturised piece of equipment. The variety of signal processing techniques available in ASP design is wider than many people realise, and some of these techniques are illustrated in the rest of this article.

Bounding, limiting and clipping

It is frequently necessary to limit the maximum excursion of a signal, for example when it is contaminated by large spikes of interference, or when the lower amplitude parts of it need to be amplified for more detailed measurements. Figure 8.1(a) shows the traditional way of doing this,

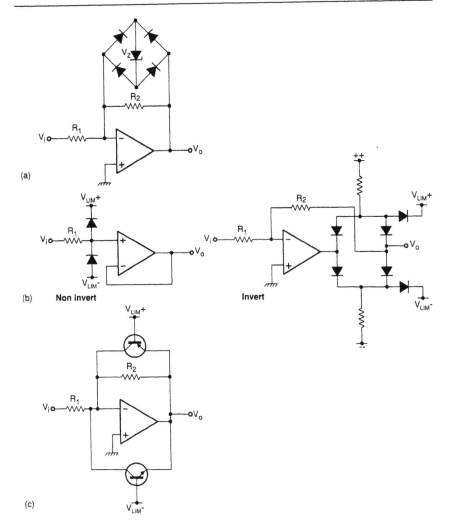

Figure 8.1 *(a)Circuit providing symmetrical limiting in an inverting amplifier. (b) Two circuits providing separately adjustable positive- and negative-limiting levels – though the inverting version is not recommended. (c) Transistors provide very effective limiting in an inverting circuit.*

employing an inverting op-amp whose gain falls from R_2/R_1 to R_s/R_1, where R_s is the slope resistance of the zener diode and the bridge diodes, whenever the output tries to exceed $\pm(V_z + 2V_s)$, where V_z is the breakdown voltage of the zener diode and V_s is the forward voltage of the bridge diodes. Thus for large signals, the gain falls almost to zero, or minus infinity dB. Disadvantages of the circuit are possible loss of bandwidth due to the capacitance associated with the bridge circuit shunting R_2, and the fact that the positive and negative limits are not easily and separately adjustable, being fixed at the same value.

Figure 8.1(b) shows a simple non-inverting circuit with separately adjustable positive and negative breakpoints V_{lim+} and V_{lim-}, and a distinctly more complicated inverting version. The latter is not recommended for fast signals, since in the overdriven condition the op-amp is left open loop, so that its output will fly off and hit one or other of the supply rails. Recovery of a conventional op-amp from overdrive is a relatively slow process, limiting the bandwidth of the circuit. If you can live with this, it is simpler to go for a larger gain and simply let limiting occur at the rail voltage. Transistors provide very effective limiting in an inverting circuit, provided the output swing keeps within the reverse V_{be} ratings of the devices. V_{lim+} and V_{lim-} are separately adjustable, Figure 8.1(c).

Another example of limiting is a circuit designed to measure the settling time of an op-amp, by using a 'false sum node', Figure 8.2(a). Low capacitance Schottky diodes, with their low forward voltage drop, are recommended in this application, enabling the 1% settling time to be measured. In the case of a fast op-amp, the capacitive loading of a conventional ×10 divider probe on the false sum node may limit measurement accuracy, so an active probe may be substituted. The measurement also depends critically upon the flatness of the top and bottom of the test squarewave. So for measurements of settling time to the 0.1% or 0.01% level, more sophisticated limiting arrangements are called for, see[1].

Where a wideband amplifier with symmetrical limiting is needed, the Linear Technology LT1194 with its 35MHz −3dB bandwidth provides a simple and convenient solution. This ingenious device[2] provides a limiting level adjustable by means of a control voltage V_c in the range −5 to −1V, with no requirement for any additional components whatever, see Figure 8.2(b). An additional advantage of this unique device is that the gain-defining negative feedback loop is completed via a second long-tailed pair in parallel with the input pair. Thus the inverting and non-inverting inputs are effectively floating and both present a high input impedance.

The circuit of Figure 8.1(b) provides protection of the op-amp's input circuit, whereas some of the others only limit the output swing. Where a circuit needs to be protected against really large inputs, the arrangement of Figure 8.2(c) can be used. For large positive inputs, D_1 is reverse biased, whilst for large negative inputs D_2 is reverse biased, the output voltage being limited to about ±5V. Given suitable reverse voltage ratings for D_1 and D_2 and a large enough dissipation rating in R_1, the arrangement will protect any following circuitry from connection (accidental or otherwise) to 230V AC mains. In the linear range, inputs between +5V and −5V, V_o follows V_i but with a possible offset due to any difference in the forward voltage drop in diodes D_1 and D_2.

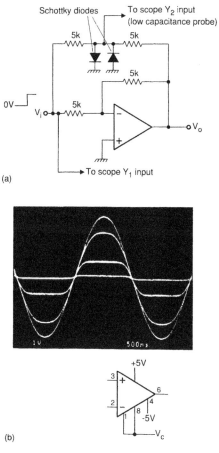

(a)

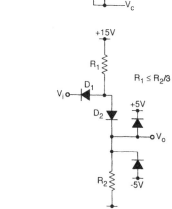

(b)

Figure 8.2 *(a) Diode limiting used in a circuit to measure the settling time of an op-amp. (b) Wideband op-amp circuit with voltage-controlled symmetrical clipping level. (Reproduced by courtesy of Linear Technology Corporation.) (c) Simple bounding circuit protects following stages from mains inputs.*

Clearly, with ±15V supplies for R_1 and R_2, if R_1 were only equal to $(R_2)/2$, it could only just succeed in raising the output to +5V, leaving no spare current to charge the inevitable stray capacitance up rapidly. To avoid a poor frequency response, a lower value, such as one third of R_2, is recomended, even though this increases the dissipation in R_1 for large negative input voltages. If the circuit is turned upside down and the diodes all reversed, an N channel MOSFET with a 600V drain voltage rating, operating as a constant-current generator can be substituted for R_1. Now, the dissipation in this 'active R_1', with a large input voltage, is only proportional to the input voltage E, not to E squared.

Breakpoints and non-linear gain

Figure 8.1(a) is an extreme example of a circuit with non-linear gain. But there is often a requirement for the gain of a circuit to vary over a range of finite values as the output level varies, rather than suddenly falling to zero. This can be arranged in many ways. The circuit of Figure 8.3(a) provides increasing gain as the input increases in the negative direction, since initially R_1, R_2, etc. are in parallel with R_B, but their effect is successively removed as each breakpoint is exceeded. In the circuit of Figure 8.3(b), the gain decreases as the output voltage exceeds each successive breakpoint, as additional feedback resistors are added in parallel with R_B. With reversed diodes (or NPN transistors) and negative breakpoint voltages, operation is extended to negative-going outputs for both circuits. Both types of break-point may be used together to give more complicated shaping as was done to linearise the frequency base in Ref. 3.

The circuits of Figures 8.3(a) and (b) will provide a gentle transition from one slope to the next, extending over a range of around 100mV or so, as the diodes or base/emitter junctions move from cut-off to conducting. Despite some consequent variation in breakpoint with temperature, this rounding can often be beneficial. However, where sharply defined break-points free from temperature variations are mandatory, the circuit of Figure 8.3(c) can be used. Here, the diode drops are all within the loop and so do not affect circuit performance. When this circuit first appeared, the use of one op-amp per breakpoint was considered almost profligate, but nowadays high performance quad op-amps are a commodity product.

Smoothly varying gain without discrete breakpoints can be achieved with various arrangements employing JFETs, either as elements in feed-back networks, or to vary the control voltage of a VCA – voltage controlled amplifier. An example of a smoothly varying function, of a very specific nature, is shown in Figure 8.3(d). Here, use is made of the inherent law governing PN junctions to provide an output voltage which is proportional to the logarithm of the input voltage or current over a wide range, up to

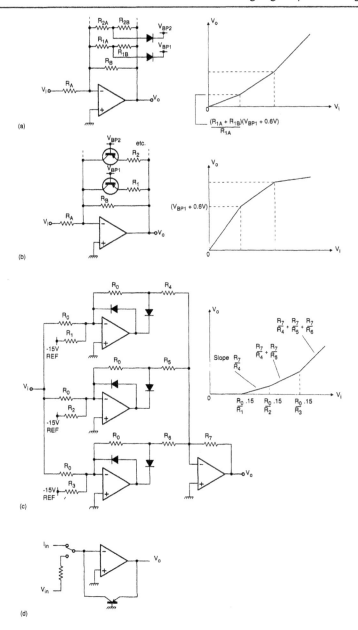

Figure 8.3 *(a) Diode breakpoint circuit, providing increasing gain with increasing output voltage. (b) Transistor breakpoint circuit, providing decreasing gain with increasing output voltage. (c) Circuit providing sharp, temperature-independent breakpoints. (d) Circuit providing an output voltage proportional to the logarithm of input voltage or current.*

nine decades with suitable devices. An op-amp with high open loop gain is necessary, to keep the base collector voltage very close to zero, otherwise collector leakage current ruins the log law at the lowest input levels. The circuit also becomes very slow at very low input levels, due the parasitic parameters of the transistor.

Slew-rate limiting

A slew-rate limiting circuit can be used to prevent the rate of change dV/dt of a signal exceeding some design maximum, whatever the actual amplitude of the signal may be. This is often necessary in electronically controlled mechanical systems with large inertia, to prevent excessive forces being applied to moving parts.

A 'leaky integrator' can form a simple slew-rate limiting circuit, as shown in Figure 8.4 a. The circuit in (ii) is obviously functionally identical to that in (i). In both cases, the slew rate is limited by the feedback via the capacitor. If now a degree of gain is incorporated in the second op-amp stage as in (iii), the full output swing will be obtained with only a reduced swing appearing across the capacitor. This is equivalent to reducing the value of the capacitor, changing the frequency at which the stage's frequency response starts to roll off, without changing the low frequency gain. So varying the amount of gain in the second op-amp stage provides a variable slew-rate limit as illustrated in Figure 8.4(b). The arrangement is simply a linear amplifier with a high frequency roll-off. Consequently, for a fixed setting of the pot, increasing signal amplitude results in increased slew rate. Thus the circuit can set any desired limit to the maximum slew rate of the *largest* signal, but as the signal gets smaller so does the slew rate, see Figure 8.4(c). The largest signal input can be set by means of a limiter, such as that shown in Figure 8.4(d). This circuit provides unity gain for small signals, but if (for example) the pot is set midway, the maximum output swing will be limited to just one half of the op-amp's rail-to-rail capability, and progressively less as the wiper of the pot approaches ground.

If the integrator of Figure 8.4(a) is placed second instead of first and a few other changes made, an improved slew-rate limiter results; see Figure 8.5(a). To understand how the circuit works, imagine that the wiper of the pot is at the top of its travel, so that R_2 and R_3 are in parallel, defining the gain of A_1 as $\times 18$, inverting. Figure 8.5(b) shows the output of A_1, upper trace, and of A_2, lower trace, when a 300Hz squarewave is applied, of just sufficient amplitude to provide the maximum output swing of which A_1 is capable. Initially, A_1 works as an inverting amplifier, because the voltage across C cannot change instantaneously. Thus the negative-going edge of the input causes the positive output at A_1, which in turn is applied via R_6 to the integrator, causing its output to slew negatively. The circuit settles with

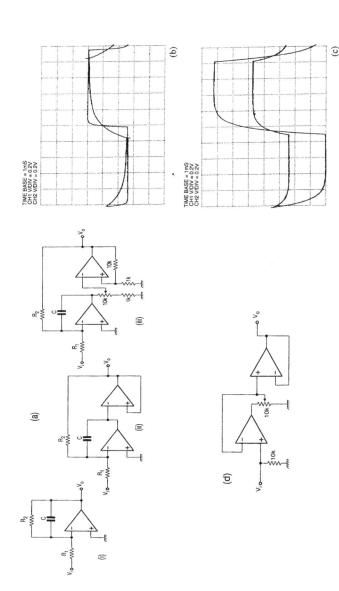

Figure 8.4 (a) A 'leaky integrator' forms a simple slew-rate limiting circuit (i), this circuit is obviously functionally identical (ii), adding adjustable gain in the second op-amp stage provides a variable slew rate limit as illustrated in (b) (iii). (b) As the gain in the second stage is increased, the required voltage excursion across the capacitance is reduced. This is equivalent to reducing its capacitance, increasing the slew rate. (c) The arrangement is a linear amplifier with a high frequency roll-off. Consequently, for a fixed setting of the pot, increasing signal amplitude results in increased slew rate. The circuit can set any desired limit to the maximum slew rate of the largest signal but as the signal gets smaller, so does the slew rate. (d) This limiter circuit can be used to define the maximum size signal input to the slew-rate limiter.

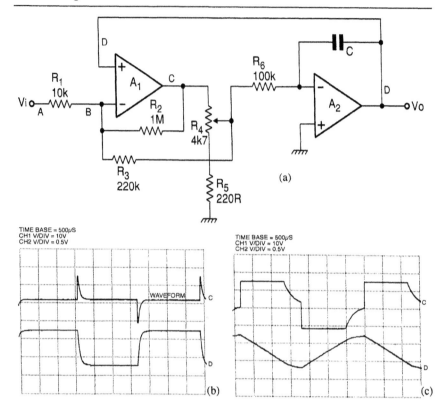

Figure 8.5 *(a) This slew-rate limiter is an improvement on the leaky integrator of Figure 8.4, providing a constant maximum slew-rate limit, regardless of the signal amplitude. (b) Set for the fastest slew-rate (wiper of R_4 at A_1's output) and with the largest signal it can handle linearly, the circuit rapidly settles exponentially. (c) With the same input, but R_4's wiper now at the junction with R_5, the circuit settles with a linear ramp, topped off with a slower exponential tail.*

A_1 output at zero (otherwise the integrator output would still be changing), and with the output voltage and A_1's non-inverting input at almost the input voltage. The exact output voltage V_o is given by $V_o = V_{in}R/(R + 10K)$, where R is A_1's effective feedback resistor – in this case, R_2 and R_3 in parallel. As Figure 8.5(b) shows, the output settles exponentially to the peak value of the squarewave input.

The case when the wiper of R_4 is wound down to the R_5 end of its travel is very different, and is illustrated in Figure 8.5(c). Now, only one twentieth of A_1's output is applied to R_3, whose effective value as a feedback resistor is therefore not 220K, but 4M4, giving in conjunction with R_2 a demanded A_1 gain of ×81. With the same input amplitude as before, A_1 is now heavily overdriven, and moreover the voltage driving the integrator stage is also reduced to one twentieth. So A_1 remains overdriven whilst the integrator output slews at a constant rate, until the voltage at A_1's non-inverting input is so near at its inverting input that A_1 re-enters the linear range. Thereafter, the circuit settles exponentially as in (b), but on a longer time-constant. So long, in fact, that in (c) the output never quite reaches the peak value before the next edge of the squarewave arrives, though if either the frequency or the amplitude of the input squarewave were reduced, it would, of course.

Whatever the frequency, amplitude or waveshape of the input, the slew-rate set by the position of R_4 is never exceeded, whilst as long as A_1 is overdriven, the first part of the settling will be at the maximum slew-rate, however small the input signal. This is a big advance on the Figure 8.4 circuit, but the exponential tail to the settling time, so visible in Figure 8.5(c) remains a disadvantage. A substantial improvement in this respect (at the cost of a reduced range of slew-rate adjustment) is obtained by connecting a $1\text{M}\Omega$ resistor between the output of A_1 and the inverting input of A_2, but though this speeds up the exponential end of the settling tail, it can never be entirely eliminated. An ideal slew-rate limiter would at all times slew at the same rate as the input signal, or at the maximum rate, whichever was the lower.

Figure 8.6(a) shows the circuit of such a true slew-rate limiter. You can see in 8.6(b) how the output follows the sinewave input from the peak (where the slope dV/dt is zero) up to the preset slew-rate limit. Thereafter, the set slew-rate applies until the loop is again closed, where the output rejoins the ideal waveform, just before the next peak. If either the frequency or the amplitude is reduced (decreasing the maximum slew rate of the signal) or the control voltage V_c is increased (increasing the OTA's maximum transconductance), the sinewave is undistorted. On the other hand, if the amplitude or frequency are sufficiently increased (or V_c reduced) the sinewave is permanently in slew-rate limit – it becomes a triangular wave.

Figure 8.6(c) shows how as the amplitude of a squarewave input decreases, the slew rate remains constant, unlike the circuit of Figure 8.4. Also, unlike the circuits of Figures 8.4 and 8.5, for any given slew-rate limit setting, the slew rate remains constant until the output rejoins the input squarewave, Figure 8.6(d). During the slew-rate limited section of the output waveform, the amplifier is open loop. Thus the OTA inverting input during this period is not a virtual earth, and exhibits the waveform shown in Figure 8.6(e). Note the pulldown resistor R_3 at the output of the operational transconductance amplifier's Darlington output buffer stage. This is recommended if using the LM13600, as the internal biasing of the buffer in this OTA is varied in sympathy with the control voltage V_c: the LM13700, with its fixed buffer current, might be a better choice in this application.

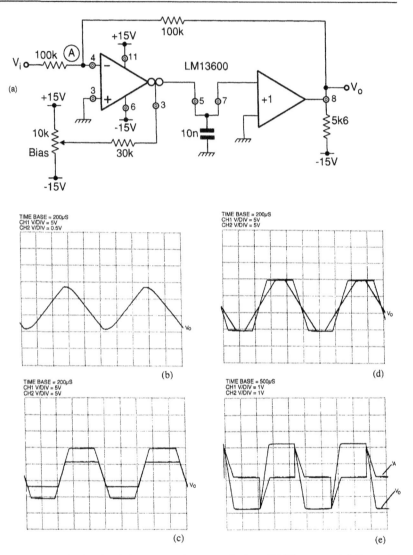

Figure 8.6   *(a) This circuit is a true slew rate limiter. (b) The output follows the sinewave input from the peak (where the slope dV/dt is zero) up to the preset slew-rate limit. Thereafter, the set slew-rate applies until the loop is again closed, where the output rejoins the ideal waveform, just before the next peak. (c) Unlike the circuit of Figure 8.4, as the amplitude of a squarewave input decreases, the slew rate remains constant. (d) Unlike the circuits of Figures 8.4 and 8.5, for any given slew-rate limit setting, the slew rate remains constant until the output rejoins the input squarewave. (e) During the slew-rate limited section of the output waveform, the amplifier is open loop. Thus the OTA inverting input (trace(a) during this period is not a virtual earth.*

Integrating and differentiating

The basic op-amp circuits for integrating and differentiating are so well known that I won't spend any time on them here, but the Howland Current Pump[4] is perhaps a circuit that deserves to be better known, Figure 8.7(a). It is a voltage-controlled current generator with (ideally) infinite output impedance, and causes a current $(V_2 - V_1)/R_1$ to flow in a load to which it is connected. If the load is a capacitor and V_1 is tied to ground, then the circuit forms an integrator. As such, it possesses two advantages over the more usual op-amp integrator: firstly, it is a non-inverting integrator, and secondly, one end of the integrator capacitor is grounded. Figure 8.7(b) shows a linear ramp or time-base generator based on the circuit.

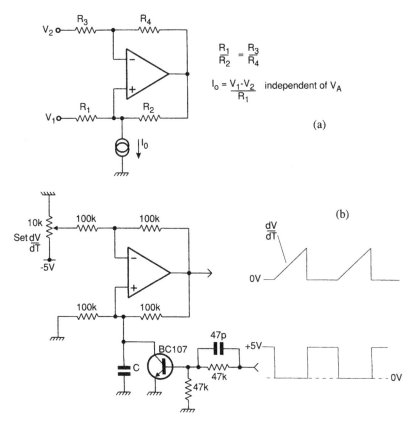

$$\frac{R_1}{R_2} = \frac{R_3}{R_4}$$

$I_o = \dfrac{V_1 - V_2}{R_1}$ independent of V_A

(a)

(b)

Figure 8.7 *(a) The Howland Current Pump provides an accurate, linear, bipolar, voltage-controlled current source. (b) One application is as a non-inverting integrator, here arranged as a linear time-base circuit.*

Some of the circuit techniques mentioned earlier can usefully be combined with integrators and differentiators. For instance, if a band-limited signal is differentiated, clipped and then integrated, the result is slew-rate limited. If instead of clipping, the differentiated signal is slew-rate limited and then integrated, then it is its second derivative d^2v/dt^2 which is limited.

Filtering

Low pass

Signals are often low-pass filtered, for a variety of reasons. Often, all the signal components of interest lie below a particular known frequency. A low-pass filter can be used to reject any unwanted signals or noise at higher frequencies. If preservation of the detailed shape of the wanted waveform is important, then some care in choice of filter type is important. For any given order of filter, a Butterworth type will give faster cut-off above the pass-band, than a Bessel filter, but it will cause ringing on a pulse type of waveform. Figure 8.8(a) compares the response of 8-pole Bessel and Butterworth filters, both with a 10kHz cut-off frequency, to a 3kHz squarewave input.

High pass

High-pass filters can improve the wanted signal-to-noise ratio in the same way as low-pass filters. For instance, a high-pass filter can be used to reject mains hum, and its harmonics also, if a higher cut-off frequency is acceptable. However, whereas a suitable low-pass filter type such as Bessel will not cause any ringing or overshoot on a squarewave or step function, overshoot seems to be an inherent feature of all high-pass filters of order higher than the first. Figure 8.8(b) shows a high-pass CR circuit with a step function input applied, giving the well-known decaying output, down to just 37% (100/e) of the input after a time interval equal to CR seconds. A series of similar sections are cascaded after the first, giving a high-pass filter with five coincident poles on the real (negative sigma) axis. This is not a good filter design, having an even 'softer' cut-off than a 5-pole Bessel, being in fact 15dB down at the 'cutoff' frequency of $f_c = 1/(2\pi CR)$, but it aptly illustrates the point. Figure 8.8(c) shows the response at the output of each stage. As initially the input to the second-stage is positive, its output will be decaying towards zero, opening up an increasing gap between second stage input and output voltages, as shown. When the output voltage of the second stage reaches zero, its input voltage is still falling, causing the output voltage to reverse sign, eventually dying away to zero from *below* the axis. A similar argument applies to the third stage, except that as the second stage

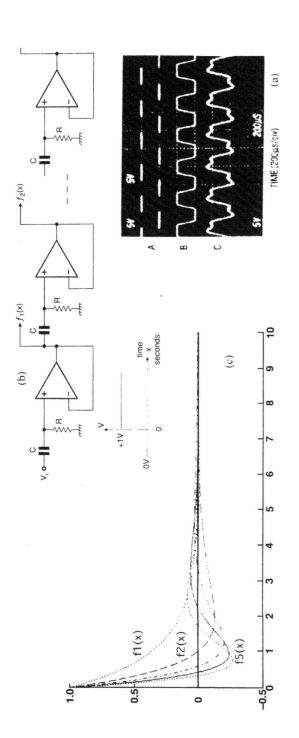

Figure 8.8 *A 3kHz squarewave A emerges from a 10kHz bandwidth 8-pole Bessel low-pass filter with its edges slowed down B, but otherwise unscathed. By comparison, a similar filter with a Butterworth characteristic causes a marked degree of ringing, C. This would be even more pronounced with a Chebychev or elliptic (Caur) filter. (b) A 5-pole high-pass filter consisting of five cascaded first order stages, with a step function input. (c) The response at each of the five outputs, showing that the nth output crosses the zero voltage baseline (n − 1) times.*

output finally dies away from a negative value, the third-stage output is pushed back across the baseline, dying away from a smaller positive value. If Figure 8.8(c) were replotted to an expanded vertical, it would be more easlily seen that the nth-stage output crosses the zero voltage baseline $(n - 1)$ times. The coincident pole high-pass filter is a minimum phase type, but if you go to a non-minimum phase design, you can have a high-pass filter with a squarer, sharper cut-off than Bessel, but still with constant group delay. The original design of such filters was for bandpass applications,[5] but this is extended to highpass types.[6]

An interesting alternative to the high-pass filter is the synthesis of a high-pass response by subtracting the output of a low-pass filter from the original signal. At very low frequencies where the output of the low-pass filter is identical to its input, the result is no signal, whilst at very high frequencies where the output of the low-pass filter is zero, the input signal simply constitutes the output signal. The less than satisfactory part comes in between, where the output of the low-pass filter is at an intermediate amplitude and phase shifted to boot. The result is partial cancellation, and a slow transition from the stop- to the pass-band. Nevertheless, the arrangement is attractive, as its implementation calls for a low-pass filter rather than a high-pass, giving one the choice of a much wider range of ready-made integrated filter circuits. Such a high-pass filter is illustrated in Figure 8.9, which also shows its limitations.

For an 8-pole Bessel filter, the attenuation at the cut-off frequency has risen to 3dB and the phase shift has reached 180°. Consequently the fundamental component of the squarewave coming through the low-pass path does not cancel out the same component via the flat path; indeed due to the 180° phase shift it actually adds to it, Figure 8.9(c). An 8-pole Bessel filter has constant group delay (phase shift proportional to frequency) well past the cut-off frequency, almost up to twice that frequency – in fact the data sheet for the MAX292 used in the circuit of Figure 8.9(b) shows it as completely flat to $2f_c$. Consequently, at that frequency the phase shift through the low-pass path has reached 360°, and the signal via the low-pass path is again subtracting from that via the flat path. However, the signal is by now much attenuated so the flat top of the squarewave is by now only slightly dinted, Figure 8.9(d), and quite flat at $3f_c$ and upwards.

The simple high-pass filter of Figure 8.9 has limited performance at low frequencies also. Figure 8.10(a) shows a 100Hz squarewave applied to the basic low-pass filter (upper trace) and its output (lower trace). The Bessel response shows no ringing, but there is clearly a finite time delay through the filter. Due to the time delay, subtracting the two waveforms allows the edges of the input through to the output before the low-pass filter output arrives to cancel the rest of each half cycle of the squarewave, Figure 8.10(b), which shows the output of the high-pass filter of Figure 8.9(b).

Figure 8.10(c) shows the effect of inserting a pure frequency independent

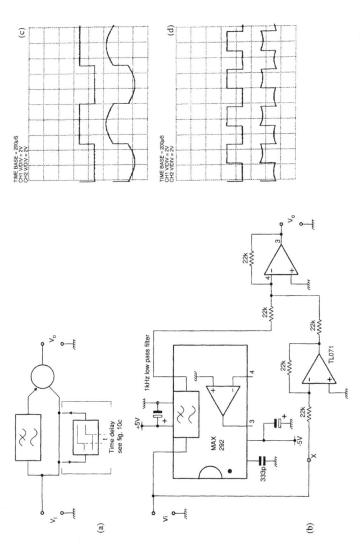

Figure 8.9 (a) High-pass response implemented with a low-pass filter. (b) Circuit diagram of a filter as in (a). The low-pass filter used was a MAX292 8-pole Bessel filter using its internal clock generator, set to give a 1kHz cut-off frequency. (c) A 1kHz squarewave applied to the high-pass filter of (b). Far from cancelling out the fundamental of the squarewave via the flat path, the component via the low-pass filter actually adds to it somewhat, since the phase shift through the filter is 180°. (d) By 2kHz, the phase shift is 360°, but there is little signal left via the low-pass path, so the overall output is nearly square.

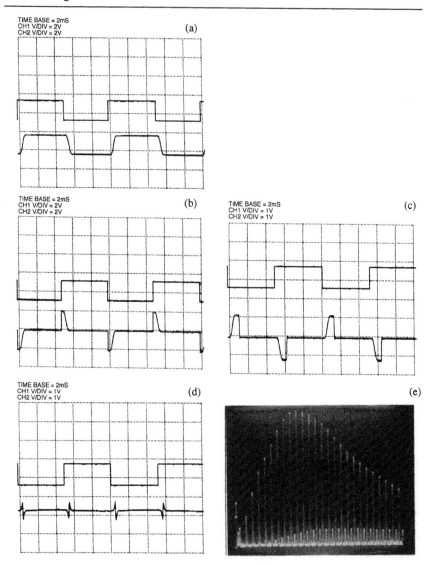

Figure 8.10 *(a) A 100Hz squarewave applied to the basic low-pass filter of Figure 8.9(a) (upper trace) and its output (lower trace). The Bessel response shows no ringing, but there is a finite time delay through the filter. (b) Due to the time delay, the simple high-pass scheme of Figure 8.9(a) and (b) lets the edges of the input through before the low-pass filter output arrives to cancel the rest of each half cycle of the squarewave. (c) Adding a time delay in the flat path – in this case excessive, as the low-pass filter output now leads the squarewave coming through the flat path. (d) With just the right delay, the direct and low-passed signals arrive at the output at the same time.*

time delay (implemented using a BBD – a bucket brigade device – of which more later) into the flat path at the point labelled X in Figure 8.9(b). In this case the BBD clock frequency was too low, giving too long a delay, with the result that the Bessel low-pass filter output now leads the broadband signal via the flat path – compare Figures 10(b) and (c). Increasing the BBD clock frequency just sufficiently reduced the delay to site the edge of the delayed squarewave in the middle of the rise time of the low-pass filter output, Figure 8.10(d). Whereas the lower trace in (b) clearly contains a component at 100Hz (together with harmonics of that frequency), the lower trace in (d) is virtually devoid of any 100Hz component. This is confirmed by the spectrum of the lower trace of Figure 8.10(d), which is shown in (e). The spectrum covers 0–5000Hz, and as can be seen, at frequencies above about 2kHz, the harmonics of the squarewave are rolling off at the rate expected. One can work back from the display to estimate the attenuation, bearing in mind that the 31st harmonic at just over 3kHz reaches up to 8 divisions on the linear vertical scale. The amplitude of the nth harmonic of a squarewave (n being odd) is $1/n$ times that of the fundamental, so here the fundamental was 31 times that of the said harmonic. The fundamental would therefore have been about 240 divisions, whereas it actually measures 1.5 divisions, giving an attenuation at 100Hz of well over 40dB.

Comb filters

It may happen that interfering signals lie within the bandwidth occupied by a wanted signal, a typical example being mains hum and its harmonics. All is not lost; it may be possible to remove the interference whilst keeping most if not all of the signal energy. The method is to use a comb filter – a filter with a series of notches at nf_0, where f_0 is the frequency of the lowest notch, and $n = 1, 2, 3$, etc.

A useful device in this context is the BBD, mentioned earlier, which subjects a signal to a time delay, the value of which is determined by the frequency of a clock signal applied to the device. Having experimented with such a device in the past, to produce artificial reverberation in connection with an electronic organ, I unearthed the breadboard, the circuit of which is shown in Figure 8.11(a). As shown, the circuit ran at the very low clock frequency of about 8kHz, limiting the useful bandwidth to about 2kHz. Thus the maximum delay provided at the output tap (stage 3328) was about 200ms, and the circuit had provision for adding in contributions from any or all of the intermediate taps, at various strengths, by means of the six-way DIL switch SW_B shown. In this device, the delays at the intermediate taps (stage 386, stage 662, etc), as a fraction of the maximum delay, are all chosen to be irrational numbers, simulating the various reflection delays and resulting eigentones in a large building.

To extend the maximum delay to one or more seconds as required, a

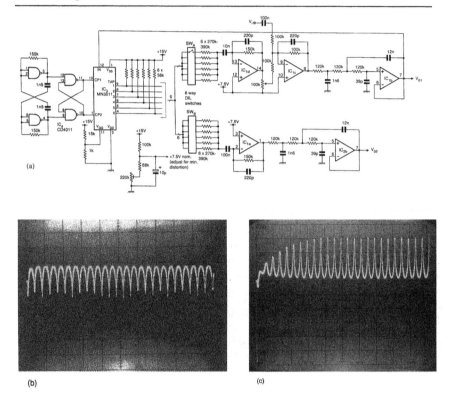

(a)

(b)   (c)

Figure 8.11 *(a) Circuit of an experimental BBD delay system, originally used for experiments with reverberation. (b) With all sections of SW_A off, and a delayed inverted version of the signal summed with the original input, a comb notch filter resulted. (Delayed signal taken from BBD tap 1, stage 386, giving a delay of 25ms. Spectrum analyser display: vertical, 10dB/div.; horizontal, 0–1000Hz span.) (c) With the pin 9 output from stage 396 selected via SW_A, the degree of recirculation was set just short of oscillation, resulting in a selective comb-pass filter. (Delay and analyser settings as (b) above.)*

similar arrangement of select-on-test resistors and DIL switch was used in the path via IC_{1d}, $_c$ and $_b$. The output V_{o1} consists of the input signal, plus selected delayed versions summed at the virtual earth input of IC_{1d} and added back in with the direct signal. (For the purposes of simulated reverberation, it was in fact found best to recirculate solely via the maximum delay, tap 3328, adding different delays only at SW_B, the input of the output path to V_{o2} via IC_{1a}, IC_{2b}.)

The MN3011 BBD chip used (R_s No. 631-294) is implemented in PMOS technology and is thus rated to operate between a V_{ss} of 0V and a V_{dd} of $-15V$, although for convenience it is usual to operate it from a positive supply, as shown in Figure 8.11(a). The chip requires a two-phase clock which is basically non-overlapping and this was provided by the CD4011, IC_4. Any overlap should be limited to within 3V of the positive rail, a condition which the circuit shown very nearly met. The circuit operates with the op-amp inputs referenced to a nominal midpoint voltage which is adjustable by means of the 220K pot, this being set for minimum distortion on maximum amplitude signals. The 100K pot at the output of IC_{1d} permits adjustment of the degree of recirculation, and hence of the length of reverberation. Naturally, if advanced too far, the loop gain exceeds one and a continuous oscillation is set up.

Comb notch filter

The BBD circuit was modified to run as a comb notch filter, by combining the signal with a delayed and inverted version of itself, but without any recirculation of the signal via IC_{1d}. The first tap, stage 396 output, was used, giving a delay of 25ms. Thus a sinewave at 40Hz would be delayed by exactly one cycle, and when inverted and added to the original signal, they should cancel each other completely. The same should apply at 80Hz, 120Hz etc., whilst at 20Hz, 60Hz, 100Hz, where the delay is 180°, the inverted delayed signal will add to the orignal, resulting in a 6dB increase in amplitude. This is illustrated in Figure 8.11(b), where the broad pass-bands are generally 20dB above the narrow stop-bands or notches. Greater notch depth can be achieved by careful adjustment of the gains in the direct and the delayed inverted channels. The reduced amplitude of the lowest frequency pass-band, centred on 20Hz, is due to the 100nF coupling capacitor at the input.

A filter with a series of narrow notches can be used to remove mains hum and its harmonics from a signal, while leaving much of the wanted signal energy. If the wanted signal happened to be at 20Hz and/or any of its odd harmonics, it would be passed by the filter in its entirety, completely unscathed. This arrangement, with a comb spacing of 10.125kHz and known as the 'tête bêche' (head to toe) system, was used years ago to distribute two black and white TV pictures over cable in urban areas. The two programmes occupied the same bandwidth, but one with its carrier at the bottom of the band and the other at the top. The carrier frequencies were set so that the line frequencies interleaved, there being negligible energy except in the immediate vicinity of each line frequency harmonic in normal picture content. Appropriate filters separated the one signal from the other, there being but two programmes to choose from in those days. Comb filters are also proposed for use with the proposed American 6MHz-channelled VSB HDTV standard signal format, for protecting the small HDTV pilot tone from interference from strong NTSC signals at the receiver.

Selective comb-pass filter

If the inverted delayed channel signal is removed and a portion of the delayed signal is added back into the original signal (by closing one of the switches in SW_A), what were previously notches now correspond to frequencies where the delayed signal is in phase with the original, giving a series of peaks. Midway between each pair of peaks, the delayed signal is in antiphase with the original, so the output is reduced. When the degree of recirculation is increased to the point where V_{out}/V_{in} at the peaks reaches (almost) infinity, the gain via the BBD and IC_{1d} must be very nearly unity, since we can almost do without the input entirely. In this condition, the gain at the intermediate points must be $\times 0.5$. This follows necessarily, since V_o equals V_{in} minus the delayed version, and the gain via the delayed path has been set to unity; the only solution to the equation $x = 1 - x$ is $x = 0.5$.

Such a selective comb-pass filter was set up using the circuit of Figure 8.11(a) by selecting the stage 396 tap output at pin 9 of IC_3 with DIL switch SW_A, the resistor from there to IC_{1d} input having been changed to 150K. With the pot at the output of IC_{1d} advanced until the circuit almost oscillated, the frequency response V_{in} to V_{out} was as in Figure 8.11(c). The uniformity of the peaks, from about 330Hz upwards, is remarkable. The roll-off below this frequency is due to the 10nF coupling capacitor at the input of IC_{1d}. (This value was selected during the earlier synthetic reverberation experiments, to avoid boominess – a realistic reverberation effect does not require an extended frequency response, in either the treble or the bass.) The effect of the 10nF capacitor on the peaks is much more pronounced than on the troughs, where the feedback via the delay is negative. This nicely illustrates how positive feedback emphasises variations, whilst negative feedback flattens them out.

A selective comb-pass filter can be used where the wanted signal and its harmonics is at a known frequency. By setting the BBD clock frequency so that the peaks pick out the wanted signal components, the system bandwidth can be substantially reduced, cutting out much of any broadband noise which may be present. However, carried to extremes, the narrow bandwidth of the peaks means that very rapid changes in the wanted signal cannot be accurately followed.

BBD rescues wanted signal

An interesting alternative is to use a selective comb-pass filter not indeed to pick out the wanted signal from the interference, but to pick out the interference from the wanted signal. The interference can then be subtracted from the original mixture of signal and interference, leaving just the signal. This application is feasible where the interference is an unvarying waveform, or in signal processing jargon, the interference is a 'stationary' signal, which means just the same thing.

Before trying out this scheme, suggested by a colleague at work, the earlier breadboard BBD circuit was modified to incorporate an MN3101, the recommended matching clock driver IC, in place of the CD4011. The MN3101 has several advantages over the earlier arrangement. Its two-phase clock outputs meet the BBD chip's requirement that the clock edges should cross within 3V of the positive rail. Additionally, it is tuned by just a single RC circuit, against the two RC networks used by the CMOS oscillator in Figure 8.11(a), a great convenience when fine tuning a highly selective filter. It also provides the gate bias voltage required by the BBD chip, disposing of the potential divider at pin 11 of the MN3011 in Figure 8.11(a). Tuning with a single resistor was going to be important in this application, where the narrow bandpass comb filter was to be tuned to an interfering signal.

The experiment demanded a wanted signal, and some interference. The choice of the latter was simple – that old enemy, mains hum. This often rears its head in low level circuits not merely as 50Hz fundamental – which could be rejected by a simple notch filter – but as harmonics of 50Hz as well, both odd and even. The even harmonics usually emanate from rectifier circuits, whilst the odd harmonics emanate from coupling with the leakage flux of mains transformers. For reasons of low first cost, trans-formers are commonly designed to operate the core up to a peak flux density verging towards saturation, resulting in a component of magnetis-ing current at the third and higher odd harmonics. The fundamental component of the flux stays mainly on the core, but the leakage flux is rich in third and higher odd harmonic components. The circuit of Figure 8.12(a) (with the point X open-circuit) was used to provide a source of fairly nasty mains hum, with even harmonic components provided by the recti-fier circuit, and odd components (including the fundamental) by the 25-turn coupling winding. This was wound around the outside of the transformer, thus coupling only with the leakage flux.

The waveform of the resulting hum signal is shown in Figure 8.12(b), upper trace, and carefully counting all the peaks, one can see that it contains components up to at least the fourth harmonic, but its spectrum – see Figure 8.12(c) – is more revealing. It shows that the fundamental, second and thirdharmonics are all at the same level (within a dB or so), whilst even the ninth harmonic is not much over 20dB down on these, and not until the 18th are the harmonics more than 40dB down on the fundamental.

The modified version of the BBD breadboard circuit of Figure 8.11(a), now incorporating the MN3101 clockdriver and intended to act as an

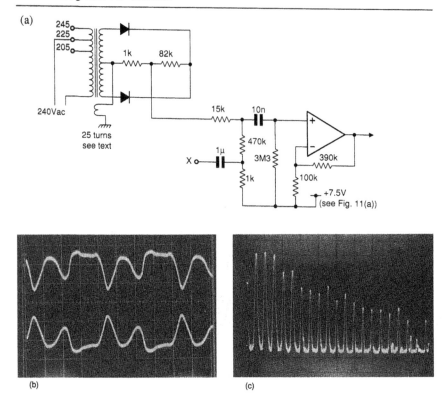

Figure 8.12 *(a) Circuit used to produce a mains-related interference waveform. (b) The resulting waveform. (c) The spectrum of the mains interference in (b). (Display: vertical, 10dB/div.; horizontal, 0–1000Hz, resolution bandwidth 3Hz, post-detection filter off.)*

interference cancelling circuit, is shown in Figure 8.13(a). The mains interference waveform of Figure 8.12(b) (upper trace) was fed in at V_i. The gain to this signal, from V_i to V_{o1}, is ×0.1 or −20dB, set by the ratio of R_1 and R_2, at IC_{1c}. However, the BBD clock frequency was carefully set so that the lowest frequency response peak was at 50Hz, and R_3 at the output of IC_{1d} then adjusted so that V_{o1} equalled V_i – apart from the inversion in IC_{1c} – as can be seen by comparing the lower trace (V_{o1}) in Figure 8.12(b) with the upper (V_i). Thus the gain around the loop IC_{1c}, IC_{1b}, IC_3, and IC_{1d} was ×0.9, and the recirculated signal, added to the smaller input signal via R_1, results in the amplitude of V_{o1} being the same as V_i.

If now V_i is disconnected, the amplitude of V_o will fall by 10% for the immediately following 20ms, as the input to IC_{1c} now consists solely of the signal coming from the delay line. For the 20ms following that, again, it will be just 90% of 90%, or 19% smaller, and so on, gradually dying away to

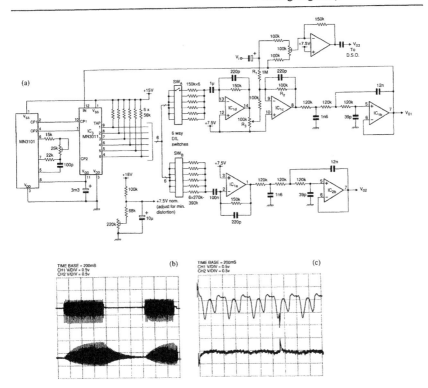

Figure 8.13 *(a) BBD device comb filter modified for use as an interference cancelling system. (b) When the interference waveform is connected to, or disconnected from, the circuit of (a), the stored replica of the interference in the recirculating system builds up or dies away gradually. (c) A 'wanted signal' consisting of occasional pips was added to the interfering waveform, top trace. When the recirculating version of the interference is subtracted from this, the wanted signal –* V_{o3} *in (a) – is left in the clear, lower trace.*

nothing. Likewise, when V_i is reconnected, initially V_{o1} will be only 10% of V_i, building up asymptotically to its full value, see Figure 8.13(b), showing V_i (upper trace) and V_{o1} (lower trace). Thus over a period of many cycles, the circuit builds up and stores a replica of the mains interference.

A 'wanted signal' was added into the mains interference, by connecting a 4V pk-pk squarewave at approximatiely 8Hz to point X in Figure 8.12(a). This was differentiated by the 1ms time constant of the 1μF and 1K CR circuit, to give narrow pips visible in Figure 8.13(c), upper trace. This was used as the input V_i to the circuit of Figure 8.13(a), which generated a replica of the mains interference at V_o, as before. The pips do not feature in this replica as they only appear infrequently in V_i, and not in the same place in the waveform each time – the pip repetition frequency is n × 20ms, where n is *not an integer*. Consequently, when the recirculating version of the interference is subtracted from V_i, the result at V_{o3} in Figure 8.13(a) is simply the wanted signal, in the clear, Figure 8.13(c), lower trace.

The recovered signal will not in fact be entirely in the clear. Firstly, its amplitude at V_{o3} will be only 90% of its input value, due summing with the the inverted $\frac{1}{10}$ amplitude version of it appearing at V_{o1}. Secondly, this inverted 10% version at V_{o1} is applied to the BBD delay input, to reappear after 20ms at V_{o1} as a reinverted 9% version. This will appear as such at V_{o3}; you might just fancy you can see it as an echo one horizontal division after the pip in the lower trace of Figure 8.13(c), lurking amongst the background noise (which is mainly clock hash).

In a practical interference suppression system using this scheme, the clock for the 3328 stage BBD chip would be maintained at exactly 2×3328 times the mains frequency, by a phase lock loop. The latter would follow any slight drift in mains frequency; as this would occur only slowly, the interference can still be considered stationary.

The scheme will work for isolated signals occurring at random intervals. It will also work for repetitive signals such as pulses (short compared with 20ms), provided that these do not recur at intervals of 20ms, or a submultiple of this. For if they do, they will appear as a stationary signal of the same period as the hum, and will build up a replica in the same way. Then, when subtracted from the original there will be nothing left.

References

1. Methods for measuring op-amp settling time. AN10-1, July 1985, reproduced in the 1990 Linear Applications Handbook from Linear Technology Corporation.
2. Hickman, I. 'Four op-amp inputs are better than two', *Electronics World and Wireless World*, May 1992, pp. 399–401, reproduced in the *Analog Circuits Cookbook*, Butterworth-Heinemann, 1995, ISBN No. 0 7506 2002 1.
3. Hickman, I. 'Add on a spectrum analyser', *Electronics World and Wireless World*, Dec. 1993, pp. 982–989, also in the *Analog Circuits Cookbook*.
4. Pease, R. A. 'Improve circuit performance with a 1-op-amp current pump, *EDN*, pp. 85–90, Jan. 20 1983.
5. Lerner, R. M. 'Band-pass filters with linear phase, *Proc. IEEE*, March 1964 pp. 249–268.
6. Delagrange, A. 'Bring Lerner filters up-to-date: Replace passive components with op-amps', *Electronic Design* 4, Feb. 15 1979 pp. 94–98.

9 Reflections on opto-electronics

Opto-electronic links are used in an increasing number of applications. This article looks at the theory behind modern opto-electronic devices, and how they are applied. It then goes on to look at practical applications, including short- (and not so short-) range data links.

Opto-electronics has come a long way since the days of vacuum photocells with caesium or silver cathodes. Nowadays semiconductor photodiodes, of silicon, gallium phosphide or GaAsP, are almost universally employed, except in certain specialised applications, such as photomultiplier tubes used in photometry. Compared to those earlier types of light sensitive cells, semiconductor photodiodes are small, inexpensive, stable and easy to use. However, there are a variety of types and some knowledge of their different characteristics is needed for their successful employment.

Silicon is the commonest material employed for photodiodes, and this material is used in various types covering from ultraviolet (UV) through the visible spectrum to the infrared (IR). Light energy falling upon the diode creates hole-electron pairs and hence gives rise to a current – if there is a circuit from anode to cathode – or an open-circuit emf otherwise. The more light energy, the larger the current, the ratio being a function of the material, and hence independent of the area of the diode.

Figure 9.1 shows the sensitivity of some typical silicon photodiodes to light as a function of wavelength, from which it is clear that the sensitivity is greatest at longer wavelengths, in the infrared, with maximum sensitivity typically being in the range 0.5–0.6A per watt. At longer wavelengths still, the sensitivity drops right off. This is because each individual photon has insufficient energy to create a hole/electron pair in the material.

The formula relating the energy of a photon E to its frequency f, is $E = hf$, where h is Planck's constant. From this it would seem evident that

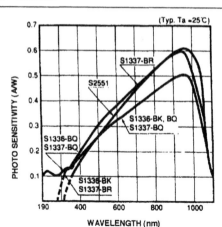

Figure 9.1   *Sensitivity of typical silicon photodiodes as a function of wavelength. (Courtesy Hamamatsu Photonics UK Ltd.)*

once the photon energy was large enough to create hole/electron pairs (i.e. in the infrared), the response should remain constant or even increase (with energetic very short wavelength photons perhaps creating more than one hole/electron pair) as the frequency increases. But, in fact, as Figure 9.1 shows, the reverse is the case. The reason becomes clear when the detailed operation of a silicon photodiode is considered.

Figure 9.2(a) shows (diagrammatically and not to scale, for clarity) the cross-section of a typical planar diffused silicon photodiode. Incident light creates hole/electron pairs. Under the influence of the potential barrier represented by the depletion layer, electrons liberated in the P layer migrate to the N layer whilst holes created in the N layer move in the opposite direction, as shown in Figure 9.2(b). This creates a current which flows through the external circuit if the diode is short-circuited, or notionally through the diode itself, establishing a voltage across it, if it is open-circuited, see Figure 9.3. Thus the diode can be represented by the equivalent circuit shown in Figure 9.3(b). When the load resistance R_L is open circuit, the illumination causes a voltage across the diode, and as with any diode this is logarithmically related to current, and shows a temperature coefficient of about $-2mV/°C$. Consequently, open-circuit operation is unsuitable for light intensity measurements. By contrast, in the short-circuit case, the current I_L (which all flows in the external circuit if the resistance of the 'short' is negligible compared to the diode series resistance R_s) is extremely linearly related to the incident light energy. When this is in the range $10^{-12}W$ to $10^{-3}W$, the achievable range of linearity is greater than eight orders of magnitude (being limited at the bottom end by noise), depending on the type of photodiode and its operating circuit.

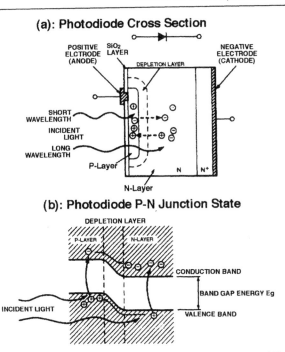

(a): Photodiode Cross Section

(b): Photodiode P-N Junction State

Figure 9.2 *(a) Diagrammatic cross-section of a typical planar diffused silicon photodiode. (b) Bandgap diagram for (a). (Courtesy Hamamatsu Photonics UK Ltd.)*

The above description applies to a basic planar diffused silicon photo diode, but many variations are possible upon the theme, including low junction capacitance types, and PIN photodiodes for high-speed response (operated with a reverse bias of up to 100V, depending upon type). Another high-speed type is the silicon avalanche photodiode, also operated with a reverse bias voltage. The avalanche multiplication effect provides internal gain in the diode itself, making it a sort of solid state analogy of a photomultiplier tube. The high value of I_L due to the internal gain (up to ×100) enables the diode to be used in a Figure 9.4(a) type circuit, with a much lower value of R_L compared to a normal photodiode. Small area silicon avalanche diodes, operated with $R_L = 50\Omega$, can achieve a cut-off frequency in excess of 1GHz, due to a low junction capacitance of around 2pF. Other types include Schottky junction photodiodes fabricated in GaP or GaAsP, offering high sensitivity well into the UV region.

Returning to the variation of sensitivity with wavelength, Figure 9.2(a) indicates that longer wavelength radiation penetrates further into the material than shorter wavelength radiation, due to absorption. The shorter the wavelength, the greater the degree of absorption of light within the surface diffusion layer, leading to reduced sensitivity, since most photons do not reach the depletion layer. In silicon photodiodes with enhanced UV

V-I Characteristics

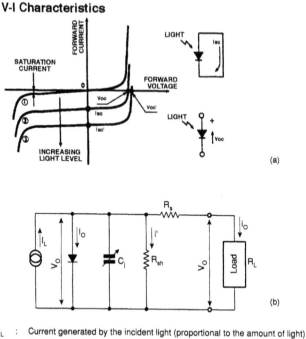

I_L : Current generated by the incident light (proportional to the amount of light)
I_D : Diode current
C_j : Junction capacitance
R_sh : Shunt resistance
R_s : Series resistance
i' : Shunt resistance current
V_D : Voltage across the diode
i_o : Output current
V_O : Output voltage

Figure 9.3 *(a) Showing how the normal diode characteristic of a photodiode is shifted when light falls upon it. (b) Photodiode equivalent circuit. (Courtesy Hamamatsu Photonics UK Ltd.)*

sensitivity, therefore, the surface diffusion layer is made very thin (the depletion layer very close to the surface).

Figure 9.4 shows two ways of operating silicon photodiodes in the current measuring mode, i.e. with and without reverse bias. In Figure 9.4(a), the reverse bias results in high-speed response to light pulses, making the arrangement attractive for high-speed data links. On the down side, linearity is poorer, noise greater and leakage via R_{sh} (see Figure 9.3(a)) results in a dark current in the absence of illumination, although R_{sh} is typically in the giga-ohm range. The arrangement of Figure 9.4(b) requires no bias source, and is very commonly employed. The high gain of the op-amp ensures a near perfect virtual earth, resulting in no voltage appearing across the diode and hence no dark current. However, the op-amp's

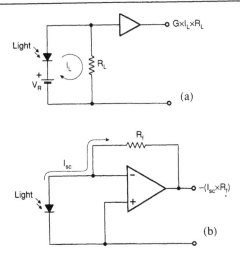

Figure 9.4 *(a) Circuit for photodiode operation with reverse bias. (b) Circuit for operation with* R_L *= short circuit. (Courtesy Hamamatsu Photonics UK Ltd.)*

bias current I_b at the inverting input has to be supplied via R_f, so if the latter is made large in order to secure high gain and thus high sensitivity, then an op-amp with very low I_b is necessary.

The sensitivity of silicon photodiodes illustrated in Figure 9.1 is the steady state or DC response, and varies as a function of wavelength (frequency) of the light as shown. However, the AC response, when the intensity of the light varies at some frequency or other, is a little more complex, due to a characteristic of silicon which has already been mentioned. The transit time of carriers liberated within the depletion layer is determined by the potential gradient therein, which in turn is set by the voltage across the layer. This may be just the band gap voltage in a Figure 9.4(b) type circuit, or an externally applied bias as in Figure 9.4(a). But carriers liberated outside the depletion region are not subject to this potential gradient and hence take much longer to diffuse to the anode, or to the cathode as the case may be.

As Figure 9.2(a) illustrates, longer wavelength radiation penetrates more deeply into the silicon, so energy in the IR may release carriers in the bulk of the N material, having passed right through the depletion layer before creating hole/electron pairs. Thus when IR illumination first strikes the diode, the current due to carriers liberated in the bulk region will appear at the terminals later than the component of current due to carriers liberated in the depletion region. At shorter wavelengths, where the luminous energy does not penetrate so deeply, this effect is much reduced or even absent entirely.

The speed of response of a photodiode, expressed as the rise-time t_r (where t_r is related to the cut-off frequency f_c by the relation $t_r = 0.35/f_c$, approximately) is determined by three factors. The first is the time constant formed by the diode terminal capacitance C_t (including the junction and package capacitance and any circuit strays) and the load resistance R_L. The second is the transit time of carriers released in the depletion region, and the third is the diffusion time of carriers released outside the depletion layer, which as noted above move much more slowly.

To investigate some of the above effects in a practical way, some experiments were carried out using a Semelabs silicon photodiode type SMP600G-EJ, the 4×4mm square silicon die having an effective area of 14.74mm^2, a responsivity at 900nm of 0.55A/W and a capacitance at 0V reverse bias of 190pF. This is mounted in a two-lead TO39 package with a standard glass window. It was connected in a Figure 9.4(b) type circuit, the actual arrangement being shown in Figure 9.5(a), using a TI internally compensated op-amp type TLE2061. In a photographic darkroom, the photodiode was illuminated by a light source modulated at about 1.2kHz, which is shown in Figure 9.5(b); for this purpose, the red LED (a 3000mcd high brightness type) being used.

Comparing Figures 9.4(b) and 9.5(a), the latter has provision for a capacitor C_f in parallel with the feedback resistor R_f. Figure 9.6(a) shows

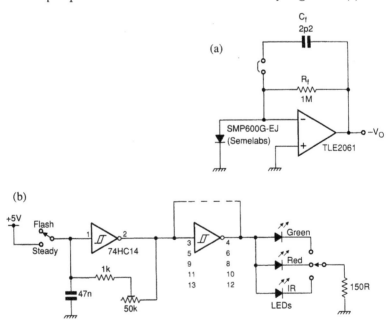

Figure 9.5   *(a) Experimental photodiode circuit. (b) Squarewave flashing light circuit used to illuminate the photodiode.*

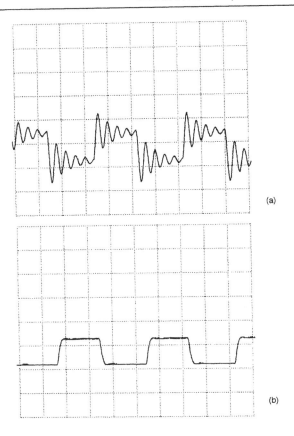

(a)

(b)

Figure 9.6 *(a) Output of the op-amp when the photodiode is receiving squarewave on/off modulated light, C_f not in circuit. (b) As (a), but C_f in circuit.*

the output of the op-amp when the photodiode was illuminated by the LED, at a suitable level, C_f not being in circuit. Severe ringing is evident.

One advantage of the Figure 9.4(b) circuit is that, as the photodiode is connected to a virtual earth, R_L is zero, and the first factor limiting the speed of photodiode response mentioned earlier – the time constant $C_t R_L$ – is apparently eliminated. Thus the speed of response should be limited only by the other two factors. But whilst the large gain of the op-amp at 0Hz ensures an ideal virtual earth in the steady state, as the frequency increases, the gain of the op-amp falls, and so a finite drive voltage is now required at its inverting input.

The op-amp's gain typically falls at 6dB/octave beyond 10Hz (e.g. in an internally compensated op-amp with a single dominant pole), and this is associated with a 90° phase lag. The passive CR circuit comprising R_f and C_t contributes another −6dB/octave roll-off and 90° phase lag at frequencies well beyond its −3dB corner frequency. If this occurs (as it usually does) well below the op-amp's unity gain frequency, then at the frequency where the op-amp gain equals the attenuation through the CR feedback circuit, the loop gain is unity and the phase shift perilously close to 180°. The circuit will therefore exhibit a gain peak at this frequency, and the fast edges of the squarewave illumination excite this and cause the ringing observed.

The addition of C_f provides a phase advance, reducing the loop phase shift and avoiding the ringing, Figure 9.6(b), the appropriate value for the circuit of 9.5(a) being found by experiment to be 2.2pF, the rise time being about 40µs. In this case, the speed of response is limited by the characteristics of the op-amp, not by the transit time of carriers released in the depletion region, and so substituting a faster one naturally improves matters. Figure 9.7 shows the performance with the same light source and

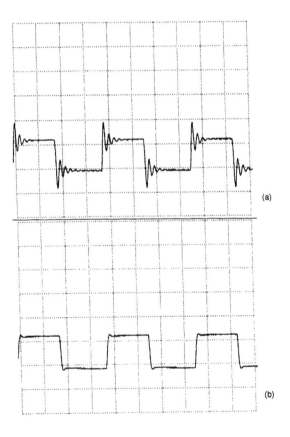

Figure 9.7 *(a) Output of the TLE2161 when the photodiode is receiving squarewave on/off modulated light, C_f not in circuit. (b) As (a), but C_f (reduced to 0.9pF) in circuit.*

photodiode, but with a decompensated version of the op-amp, a TLE2161, substituted for the TLE2061. The faster response is illustrated by the much higher frequency ringing in Figure 9.7(a) ($C_f = 0pF$) and by the reduced rise time of about 11μs in (b) ($C_f = 0.9pF$, i.e. two 1.8pF capacitors in series).

Figure 9.8 shows the effect, mentioned earlier, of carriers released outside the depletion region by long wavelength radiation which penetrates further into the material. In (a), the illumination was from chopped light from an 'ultrabright' green LED 590-345, and the edges of the waveform are square. In (b), the illumination was from an infra-red emitting GaAlAs diode type TIL901. The majority of the response is due to carriers released in the depletion region, and hence is as prompt as in (a).

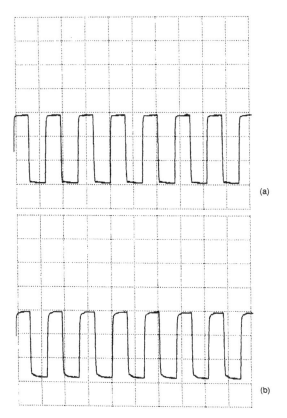

Figure 9.8 *(a) Output of the circuit of Figure 9.5(a) (using the TLE2061) when illuminated by light from a green LED (3mm ultrabright green, 590-345). (b) As (a) but receiving light from an infrared LED, TI GaAlAs IR diode TIL901. Note the delayed contribution from carriers released in the bulk material, outside the depletion region.*

But the output then rises (or falls) further, due to the much slower diffusion of those carriers released outside the depletion region.

When using silicon photodiodes, the amplifier will often be the limiting factor as far as frequency response goes. The exception is when using a photodiode (especially a low capacitance type), in a Figure 9.4(a) type circuit with a low load resistance, since here a wideband RF amplifier can be used in place of an op-amp. However, the low value of load resistance implies a relatively low sensitivity, so when detection of very low light levels is desired, a Figure 9.4(b) type circuit is used. Since designers often demand both high sensitivity and wide bandwidth, any method of extending the bandwidth of Figure 9.4(b) would be welcome. In this circuit, as the frequency rises, the op-amp gain falls, thus requiring a larger drive voltage at its inverting input – the virtual earth fails.

Instead of adding a capacitor C_f to prevent ringing, one could in principle extend bandwidth by simply adding a negative capacitance[1] of

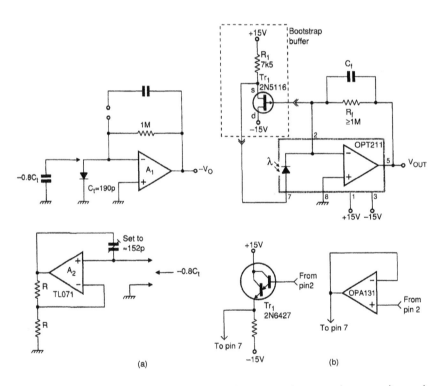

Figure 9.9   *(a) Increasing the bandwidth by connecting negative capacitance in parallel with the photodiode. (b) Increasing the bandwidth by bootstrapping, applied to an integrated photodiode/op-amp type OPT211. (Courtesy Burr-Brown Corporation.)*

$-C_t$ in parallel with the photodiode. I tried this using the circuit of Figure 9.9(a), and it does work. The snag though is that the op-amp used to provide the negative capacitance A_2 needs to have a considerably greater bandwidth than the original amplifier A_1. So if you have such an op-amp, it is simpler to use it at A_1 in the first place and forget all about negative capacitance. There is, however, a practical way to increase bandwidth – by about a factor of $\times 3$.

This is shown Figure 9.9(b), applied to an integrated photodiode/op-amp encapsulated in clear plastic, Burr-Brown type OPT211. Driving the anode of the photodiode reduces the effect of its capacitance upon circuit bandwidth. With $R_f = 1M\Omega$, $C_f = 1pF$, a bandwidth of 150kHz is achieved. Note that C_f includes the self-capacitance of the resistor R_f – a separate component may not in fact be necessary. The buffer bias current is supplied via R_f, and so should be negligible if a dark offset voltage is not acceptable. The P-channel buffer shown meets this requirement, whilst also ensuring that the anode of the diode is at ground voltage or below. The buffer bandwidth should be at least 4MHz. The two alternative buffers shown both have disadvantages, which may not be important in a given application. The Darlington buffer's bias current will result in a dark voltage offset, whilst the op-amp buffer's noise may degrade the overall noise performance slightly.

Photodiodes are often used as part of an optical signalling link, handling digital data, e.g. a TV remote control handset. Here, only the variations of incident light – the data – are of interest. Steady ambient light will produce a photodiode output which must be ignored, even if its level should change. Simple AC coupling of the photodiode op-amp's output may suffice, with if need be a sufficiently high LF cut-off to suppress 100Hz ripple due to artificial lighting. But with a high sensitivity system, where R_f is large, bright ambient light may saturate the op-amp's output. The circuit of Figure 9.10 can reject very bright ambient light, yet provide high AC gain for best signal-to-noise ratio. This is possible because of the very large linear range of a silicon photodiode. The auxiliary op-amp keeps the OPT211's mean output voltage at zero and would thus compensate not only for bright ambient light, but also for the offset due to the base current of a Darlington buffer in Figure 9.9(b).

Using a silicon photodiode in a Figure 9.4(b) type circuit, with a low noise/high gain op-amp, with ambient light rejection per Figure 9.10 if necessary, it is possible to set up optical data links subject, of course, to line of sight ranges. On a long link, the received signal will be small, and the range is therefore limited by the circuit noise of the op-amp. The total circuit noise is the sum of contributions from the op-amp's voltage noise and its current noise (both quoted in the manufacturer's data sheets), and the resistor noise originating in resistors in the input circuit. This is illustrated in Figure 9.11. In the case of a Figure 9.4(b) type circuit, R_a is

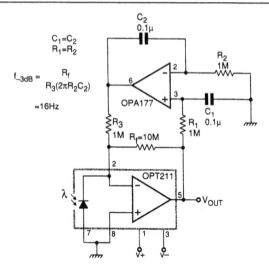

Figure 9.10   *Circuit to reject ambient light whilst providing high sensitivity to wanted signals. (Courtesy Burr-Brown Corporation.)*

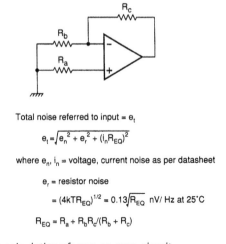

Total noise referred to input = e_t

$$e_t = \sqrt{e_n^2 + e_r^2 + (i_n R_{EQ})^2}$$

where e_n, i_n = voltage, current noise as per datasheet

e_r = resistor noise

$$= (4kTR_{EQ})^{1/2} = 0.13\sqrt{R_{EQ}} \ \text{nV}/\sqrt{\text{Hz}} \ \text{at } 25^\circ\text{C}$$

$$R_{EQ} \approx R_a + R_b R_c/(R_b + R_c)$$

Figure 9.11   *Noise calculations for an op-amp circuit.*

zero, R_c is R_f and R_b is the photodiode's shunt resistance R_{sh} (which is usually very much greater than R_f). Given these resistance values, and the data sheet figures for the op-amp's voltage and current noise, the total noise e_t referred to the input can be calculated from the formula in Figure 9.11.

An experiment was carried out to estimate the range that could be

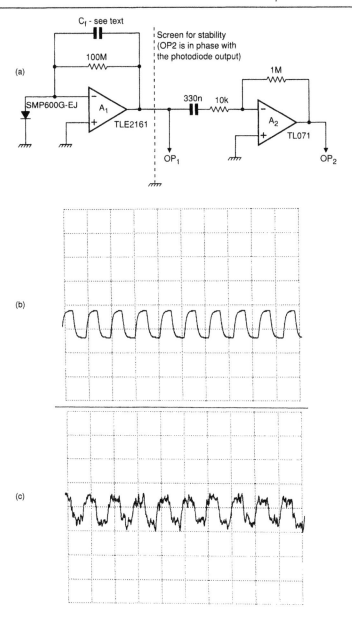

Figure 9.12 *(a) Receiving end of a long range optical data link. (b) Output of A_1 at a range of 2m from 3000mcd red LED. (c) Simulated output of A_2 at a range of 20m.*

expected for a data link of very modest data rate, using the circuit of Figure 9.12(a). In the interests of high sensitivity, the feedback resistor R_f has been raised to about 100MΩ. The necessary C_f was too small to use a discrete component, so two short pieces of insulated wire were soldered to the ends of R_f (which was actually two 47MΩ resistors in series), and twisted as necessary to just suppress ringing. The range was clearly greater than could be accommodated in the author's laboratory, so the range for a conveni- ent-sized output was measured monitoring the output of A_1 rather than A_2. Using the 3000mcd red LED, this turned out to be almost exactly 2m, the output from A_1 being shown in Figure 9.12(b). The oscilloscope probe was then transferred to the output of A_2 and the LED aimed off, so as to provide the same amplitude signal, simulating an increased path loss. As Figure 9.12(c) shows, the signal was still just usable, being adequately above the noise level.

The optical path loss follows an inverse square law, i.e. increasing by 6dB for every doubling of the range. The 100 : 1 reduction in received voltage at the output of A_1 corresponds to 100 : 1 reduction in photodiode current and thus to 100 : 1 reduction in received light energy. Given the range law, this predicts an estimated range of 20m when using the output of A_2 with the transmitting LED properly aligned. The frequency of the squarewave signal is a little under 2kHz (it was meant to be 2.4kHz) and can be taken as representing 'revs' (reversals), or a continuous pattern of alternate 0s and 1s. Thus the link would support NRZ signalling at, say, 4.8kb/s, as it stands.

This range could be extended by some simple optics. A 50mm diameter lens at the receiving end would gather one hundred times as much light energy as the photodiode alone, giving a ten times increase in range when properly aligned, whilst a similar increase could be obtained by a lens at the transmitter to reduce the 15° beamwidth of the 3000mcd red LED, at the expense of yet more critical alignment when setting up the link. With 10 000mcd types now available, another factor of almost two on the range is also possible, giving a theoretical factor of 200 increase in range, to 4km! This being more than one is likely to use, range can be traded for bandwidth by reducing R_f, permitting a higher data rate at a more practical range, such as is used in the cordless optical link between a computer and a laser printer.

References

1. Hickman, I. 'Negative approach to positive thinking', *Electronics World and Wireless World*, March 1993 pp. 258–261.

Part 2
AUDIO

10 Low distortion audio oscillator

With a total harmonic distortion plus noise output over the audio range of 0.0006% or less, the low distortion AF oscillator presented here provides a well-nigh impeccable signal source for Hi-Fi testing. Indeed, a special test rig was needed to verify the quoted performance.

There has been a long tradition of interest in the pages of *Electronic World and Wireless World* in high fidelity audio, and this of necessity includes instruments for measuring the performance of Hi-Fi amplifiers and other audio systems. Central to distortion measurements are low distortion oscillators and distortion meters, and examples of both of these have appeared from time to time in these pages.[1,2] Ref. 2 is much the earlier and only claims to be suitable for measurements down to 0.01% THD (total harmonic distortion), so it would not be suitable for use with modern equipment having levels of THD at or significantly below this. The oscillator of Ref. 1, however, with its claimed performance of <0.005% THD over the whole range 20Hz–20kHz and as low as 0.0005% typical (or even 0.0002% at 1kHz, actually lower than the commonly accepted typical distortion figures for the NE5532 op-amps used), would be very suitable for evaluating the performance of a THD measuring system as described in Ref. 2. In fact, low distortion oscillators and matching THD meters form a chicken-and-egg pair, with each being (ideally) tested with a sample of the other having a level of internal distortion much lower than its own. This aspect, and the 'because it is there' syndrome familiar to would-be climbers of Everest, has prompted me over the years to experiment with various AF oscillators and THD meters. A trawl through my files unearthed a circuit dated 17/8/1966, shown in Figure 10.1; even the required layout, using turret tagstrip, was there. It is the circuit of a THD meter which I designed and built (in an Eddystone diecast box!) and which

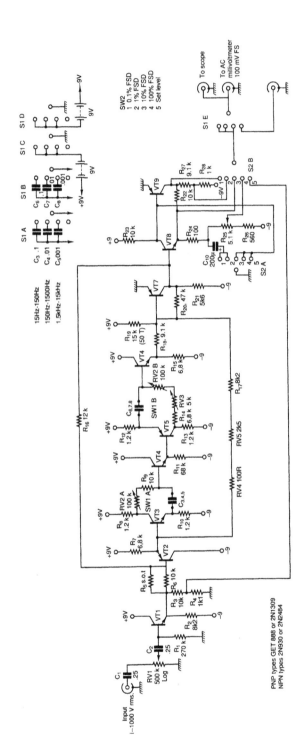

Figure 10.1 Circuit of a THD monitor designed by the author and dating from the mid sixties. Like that in Ref. 2 it was designed to be used with a separate external AF millivoltmeter.

served me and my then employer well for a number of years, though its performance was doubtless very poor by modern standards. It predates Ref. 2 by some years and was in fact suggested by an even earlier article in *Wireless World*, entitled 'The Selectoject' – an instrument which could be switched to operate as either a notch filter or a selective amplifier; unfortunately I no longer have the reference. Like Ref. 2, the distortion monitor of Figure 10.1 was designed for use with a separate external AF millivolt-meter, which would have been average responding, scaled to rms on a sinewave. Thus the indicated distortion would only approximate the true figure, see Box.

More recently, about 10 years ago, a new THD meter was designed, based around the state variable filter (SVF). This offered ranges in a 30 - 10 - 3 - 1 sequence down to 0.01% FSD, permitting 'measurements' down to 0.001% or less, although its residual noise level corresponding to 0.0009% in a 20kHz bandwidth, 0.003% in 80kHz, is deplorably high. It had a built-in true-rms responding indicating section using the well-known AD536 rms to DC converter. This instrument worked well and is still in use, although when making measurements at the 0.001% level one is demanding a degree of suppression of the fundamental of well in excess of 100dB, a tall order for a single notch. This is especially the case in a THD meter where the notch is enclosed within a negative feedback (NFB) loop to ensure that the response is back up virtually level at twice the notch frequency, where the second harmonic is encountered. The result is that any slight frequency instability in the test source, appearing as noise sidebands, will cause energy to appear peeping out either side of the notch and raising the level of the measured residual. The NFB mentioned as necessary to ensure a virtually flat response at second and higher harmonics of the rejected fundamental, also causes a bit of a design problem. The flattening of the frequency response is bought at the price of an increase in gain at the fundamental at an internal circuit node, as shown in Figure 10.2. This shows part of the THD meter of Ref. 2, based around a Wien bridge, though a very similar argument applies whatever circuit is used to imple-ment the sharpened-up notch. The level of the enhanced fundamental component must of course be kept well below clipping level, limiting the permissible input level to the notch stage. This in turn reduces the dynamic range due to the reduced clearance of the signal above the wideband noise floor, providing another factor, in addition to the instrument's own residual internal distortion, limiting the lowest level of distortion that can be observed. Nevertheless, despite this, a high performance THD meter *should* be easier to design than an oscillator, since all circuitry within the former can be linear, whereas the latter requires an amplitude control mechanism, which involves some non-linearity. The SVF-based instrument mentioned above is actually, like many THD meters, an 'everything else' meter, responding to harmonics of the input, hum and the wideband noise floor, as

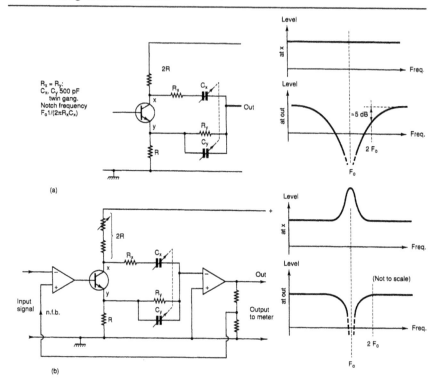

Figure 10.2   *(a) Circuit providing a notch at one frequency F. The response at 2F is still 4.77dB down, eventually returning to the DC value at a much higher frequency. (b) By enclosing the notch circuit within an NFB loop, the response can be sharpened up to be no more than a fraction of a dB down at second harmonic. However, internal to the loop, the response at the fundamental is actually peaked up significantly.*

well as any significant spectral noise sidebands surrounding the fundamental which fall outside the notch. Useful as this instrument proved, initially there was no way of knowing what was the level of residual distortion in the instrument itself, so a few months ago the development of a very low distortion oscillator was undertaken, to provide a test source.

As mentioned, some non-linearity is needed to constrain an oscillator's output amplitude at a suitable constant level. This non-linearity can operate on a cycle-by-cycle basis, as in Ref. 3, or over many cycles as in Ref. 1 where the non-linearity is provided by a thermistor. In this latter case, at a certain amplitude of oscillation the power dissipation in the thermistor is sufficient to cause its resistance to fall (it has a negative tempco) to a level where the net loop gain is exactly unity at just that signal level. Having a thermal time constant of nearer a second than a millisecond, the thermistor's resistance remains sensibly constant over each cycle of the output, except at frequencies below 100Hz. Here, and especially so down at the lowest frequency of 20Hz, it tends to heat and cool slightly on each individual half cycle, leading to 20Hz distortion figures in the range 1% or worse for a typical design, down to 0.1% in the case of Ref. 1 (reduced at the output to <0.005% by a distortion cancelling technique described in the article). Because of this rise in low frequency distortion, I decided against using thermistor stabilisation. However, experience had shown that while a cycle-by-cycle non-linearity design is very convenient for an oscillator of modest performance such as 0.05% THD, it is very difficult to design a really low distortion oscillator using this approach. It was therefore decided to use an SVF-based design incorporating non-linearity operating over many cycles, effectively adjusting the loop gain as in the Ref. 1 design. In fact, it is actually the loop phase shift which is adjusted, since with the two-integrator loop of the SVF, there is always a frequency at which the loop gain is unity. (This contrasts with the approach in Ref. 1 which uses an all-pass filter-based oscillator: here there is always a frequency at which the loop phase shift is 360°, so it is the loop gain which has to be adjusted to obtain a stable output.) The circuit adopted is shown in block diagram form in Figure 10.3(a), with the operating principle explained in Figure 10.3(b).

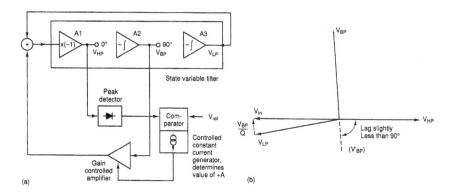

Figure 10.3 *(a) Block diagram of the SVF filter-based low distortion oscillator described. The filter selectively amplifies the fundamental component of the output of the variable gain amplifier, discriminating against any harmonic distortion present. (b) Showing how the BP output, lagging the HP by a fraction less than 90° (much exaggerated for clarity), looks as though it is leading by just over 90°, as A$_2$ is an inverting integrator. Similarly LP with respect to BP, so LP lags HP by slightly less than 180° and cannot by itself provide the necessary input to the filter V$_{in}$ to produce the output shown. Addition of a fraction 1/Q of the BP output increases the phase shift to 180° and gives a voltage equal to the required V$_{in}$, causing oscillation.*

Initial breadboard results were promising, which immediately brought me back to the chicken-and-egg problem of measurements. A piece of equipment that I had designed and built some years ago was therefore unearthed, refettled and pressed into service. As Figure 10.4 shows, this comprises a passive twin-TEE notch filter followed by a second-order Chebychev active high-pass filter, forming a fourth-order elliptic high-pass filter. The peaking of the Chebychev high-pass filter is set to compensate for the attenuation of the passive notch filter at twice the notch frequency, amounting to some 8dB relative to the 0Hz and far-out high frequency response. In the case of channel 2 (600Hz) shown, the LF roll-off of this elliptic high-pass filter also discriminates against 50Hz and its harmonics. The Chebychev high-pass filter section is followed by a low-pass filter designed, in the case of the 600Hz notch channel, to cut off beyond 3kHz. With this in circuit, THD up to and including the fifth harmonic is measured, very low levels being easily seen due to the reduction in wideband noise afforded by the low-pass section. In the case of a low distortion oscillator circuit, the design will be such that often only the lower orders of harmonics are significant. But in other cases, such as a high power amplifier using a push-pull output stage, higher order distortion

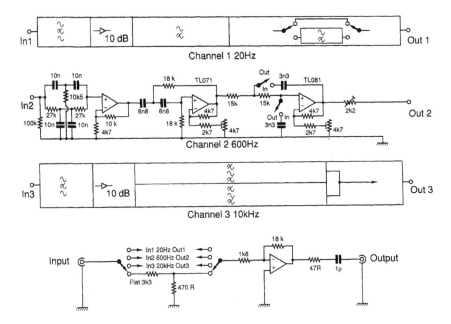

Figure 10.4   *Circuit of the 'Notch Box' used to extend the measurement range of a THD meter, providing three spot measurement frequencies of (approx.) 20Hz, 600Hz and 10kHz.*

components due to, say, crossover effects may be present. In this case, the low-pass section would be inappropriate, so provision is made to switch it out. The adjustments shown are set up so as to provide a sensibly flat response from the second harmonic of 600Hz upwards (or to the second to fifth harmonics, inclusive). The through gain in the flat position is approximately unity (actually somewhat higher), whilst it is arranged that with a notch channel in circuit, the gain over the specified range of harmonics is increased by exactly 40dB. Thus when used in conjunction with an existing THD meter, the 10% distortion range becomes 0.1% FSD, the 1% range becomes 0.01%, etc. Clearly, when making measurements using a notch channel, the effect of any internal noise or distortion in an associated THD meter is reduced by a factor of 100, whilst due to the protection afforded by the passive twin-TEE notch there should be a negligible contribution to distortion from the notch channel itself, though it will of course contribute its own noise. The 20Hz channel uses exactly the same circuit arrangement as the 600Hz channel, again with the option of switching the low-pass section in or out as desired. However, the 10kHz channel is somewhat different. The twin-TEE section is followed like the others by a buffer/10dB gain stage, but this drives two state variable filters, one tuned to 20kHz and the other to 30kHz. The outputs of these are combined in the output buffer amplifier. Clearly, then, with this 'notch box' in use ahead of a conventional THD meter, the measurement is no longer an 'everything else' measurement, especially with the low-pass section in circuit (channel 1 or 2) and particularly in the case of channel 3. Nevertheless, where harmonic distortion is the prime interest as in the present case, the arrangement is very convenient in use. In particular, tuning out the fundamental is greatly eased, as rejection is provided by the notch filter in the associated THD meter, by the passive twin-TEE notch (which is effectively wider, not being enclosed in an NFB loop to sharpen it up), and with a further contribution of around 12dB from the notch box's Chebychev high-pass filter. The great disadvantage is that measurements are confined to the three test frequencies catered for. Of course, additional channels could be provided, but the added compexity rapidly becomes cumbersome.

With means to hand to see distortion levels down to below −115dB relative to the fundamental (<0.00018%) work on the development of the low distortion oscillator proceeded rapidly. When working at very low levels of distortion, layout becomes just as important as it is in an RF circuit, so breadboarding was abandoned in favour of a form of construction which, whilst still based on discrete wiring on a commercial prototyping PCB, would result in a usable final instrument. The board was mounted in the final case, together with front panel controls and connectors, though at this stage powered by + and −15V supplies from a lab bench power supply (a linear type, not a switcher for obvious reasons). The

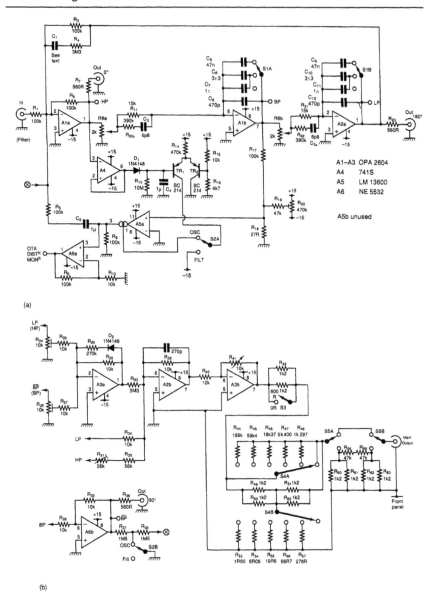

(a)

(b)

Figure 10.5   *(a) & (b)*

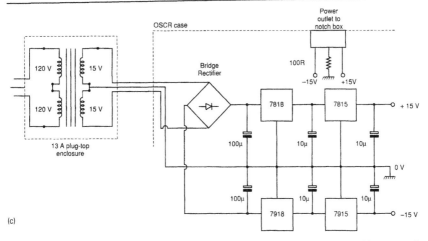

Figure 10.5 *(a) Part circuit of the SVF filter-based low distortion oscillator: oscil-lator section. (b) Part circuit of the SVF filter-based low distortion oscillator: output section. (c) Part circuit of the SVF filter-based low distortion oscillator: power supply section, also supplying the notch box if required. As the latter draws balanced currents from the +15V and −15V rails, no current flows in the 0V rail, which is normally connected between the two instruments and any intervening equipment under test such as an amplifier, via the outers of screened leads. The 100Ω resistor in the 0V rail is simply to prevent a floating earth if the screened lead(s) are disconnected.*

intention was to use the state variable filter running at the highest practi-cable value of Q, to filter a low level low distortion sinewave drive. The SVF circuit was therefore tested running without any intentional quadra-ture feedback, either negative or positive, providing in theory a filter with an infinite Q. At low and middle frequencies, the inevitable fractional shortfall of phase shift in each integrator meant that the loop phase shift fell fractionally short of 360° and the circuit was stable. But on switching to the 2–20kHz range, at the higher frequency settings the three op-amps in the loop were beginning to contribute a mite of phase shift each, leading to oscillation when the frequency was set to 17kHz or higher. This was compensated out by providing a touch of phase advance, which turned out to be not quite so simple. A single capacitor between the low-pass (LP) output and the virtual earth of the high-pass (HP) stage will certainly stabilise the circuit, suppressing any tendency to oscillate at 17kHz or above, but the damping it provides becomes quite excessive at 20kHz, leading to a low Q at this frequency. The phase advance was therefore distributed between all three stages within the basic loop, each capacitor being provided with a series resistor to limit the phase advance as shown in the full circuit diagram, Figure 10.5. The required value of capacitance at the HP stage virtual earth, Figure 10.5(a) was so small that it was provided by two 2cm lengths of wire-wrap wire, twisted more or less as required.

Now, a small quadrature (negative damping) term was taken from the BP output via an operational transconductance amplifier (OTA) and added in at the HP stage virtual earth. When this is just sufficient to bring the loop phase shift up to 360°, the circuit will oscillate (Figure 10.3(b)). A current source provides the I_{abc} bias current required by the OTA, but it is arranged that as the amplitude of the oscillation increases beyond a certain threshold, the peak-detected voltage causes a reduction of I_{abc}, reducing the loop phase shift back to exactly 360° and stabilising the amplitude at that level. OTAs are not renowned as the lowest of low distortion devices, so the voltage applied to its non-inverting input is kept to a very low level by the 100K/27R potential divider R_{17}, R_{18}. The OTA's output current is divided between two 100K resistors, one grounded to ensure that its output voltage remains centred within its voltage compliance range, and one providing DC-blocked 'negative damping' to the HP stage virtual earth. The voltage set by the 10K/4.7K resistors R_{15} and R_{16} at the base of the PNP long-tailed pair provides a +5V reference voltage. If the detected peak voltage is more or less than this value, I_{abc} will be reduced/increased respectively, adjusting the negative damping term as required to maintain a constant amplitude of oscillation. The peak detector time constant of 1 second ($10M\Omega \times 1\mu F$) is long enough to do duty on all three frequency ranges, covering 20Hz–20kHz, though it is responsible for some rise in distortion at 20Hz. However, increasing this time constant is not without its problems, so the values shown have been retained.

To see just what Q the filter runs at, I_{abc} was reduced to zero and the value of resistance required between BP output and HP virtual earth input to just cause oscillation was determined: this value divided by the value of the resistance from LP output to HP virtual earth (R_3, 100K) gives the filter Q. The circuit oscillated over the whole 20Hz–20kHz range with quadrature feedback via $10M\Omega$, over most of the range with $20M\Omega$ and some of the range with $30M\Omega$. Thus the operating Q of the loop, considered as an SVF, is generally between 200 and 300. In an active filter operating at such a high Q, the output amplitude would normally be very sensitive to temperature and other external factors. Here, however, the narrow range of level needed to change I_{abc} from zero to maximum results in very tight amplitude control. With a gain at the fundamental of, say, ×250 and an attenuation of ×3 at third harmonic in each of the integrator stages, any third harmonic in the signal from the OTA should be reduced by a factor 2250 or 67dB (a factor of 1000 – 60dB – at second harmonic). So if the distortion introduced by the OTA can be held below 0.1%, the output distortion should be <0.0001% at second harmonic, and the third even smaller: if the op-amps in the loop have zero distortion themselves, a big 'if'. The op-amp A_{6a} with its gain of ×11 was included to provide a buffered amplified version of the OTA output, as an aid during circuit development and testing. The voltage measured here confirms the estimate of operating Q, being 0.5V pk-pk at 10kHz. This corresponds to 45mV pk-pk at the OTA output as against 10V pk-pk at the LP output, indicating a Q of 220.

The oscillator low-pass output labelled LP in Figure 10.5(a) was used to drive an output section as shown in Figure 10.5(b), being fed to the virtual earth of op-amp A_{2b}. It can be seen that there is also a contribution from the HP oscillator output, via $R_{31} + R_{35}$. This was included so that R_{31} could be adjusted to suppress any third harmonic component in the output, as described in Ref. 3. The output of A_{2b} drives A_{3b}, whose feedback resistor is variable, providing fine output level adjustment. Coarse adjustment in 10dB steps is provided by the 600Ω bridged TEE attenuator associated with S_4, and by the 0 or 50dB pad, S_5. The odd value resistors R_{44-48} and R_{53-57} were made up by paralleling standard values to get within 1%. S_3 provides a choice of a 600Ω output impedance or (with the step attenuators set to 0dB) a low impedance. This provides twice the maximum available peak to peak output voltage into 600Ω, but if the step attenuators are used, the first 10dB step will no longer be accurate. Note that great care is required in the earth routeing, as indicated in Figure 10.5(b), if the distortion is to go down pro rata with the output; screened lead should be used between the main output socket, S_5, S_4 and the main circuit board.

When set up, the oscillator provided the following performance, all measurements being taken with S_3 set to low output impedance, S_4 and S_5 to 0dB, and using the notch box.

20Hz (THD Mtr 20kHz BW, Notch box LPF out, Figure 10.6(a)	0.00062%
20Hz (THD Mtr 20kHz BW, Notch box LPF in)	0.00059%
600Hz (THD Mtr 20kHz BW, Notch box LPF out)	0.00034%
600Hz (THD Mtr 20kHz BW, Notch box LPF in)	0.00022%
10000Hz (THD Mtr 80kHz BW, Figure 10.6(b)	0.00092%
10000Hz (THD Mtr 20kHz BW, Figure 10.6(c)	0.00062%

At 600Hz, the residual distortion can be clearly seen when directly viewing the THD Meter's *residual* output on an oscilloscope, but with the 20 seconds exposure required by my home-made 'scope camera, it was totally obscured by the broadband noise when photographed. In fact, the distortion is visibly virtually pure second harmonic, the figure of 0.00034% being higher due solely to the extra noise admitted in the absence of the notch box's 3kHz low-pass filter.

I will not pretend that the above performance will automatically result from the circuit shown; a number of stages of setting up were required. Firstly, the Burr-Brown OPA2604 op-amps were not specially selected – only three samples were to hand – but they were swapped around for best results. (With a manufacturer's quoted typical THD + noise of 0.0003% at 1kHz, the OPA2604 FET input dual op-amp is not the lowest distortion

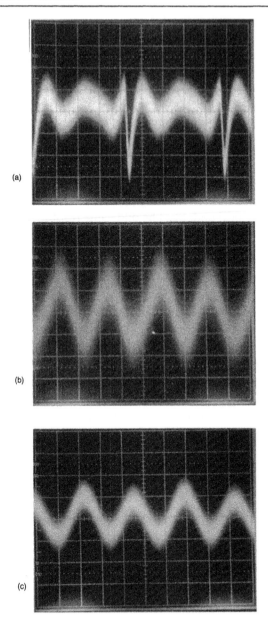

Figure 10.6   *(a) Residual distortion plus noise at 20Hz is 0.00062%. (b) Residual noise and 2nd + 3rd harmonic distortion with THD meter bandwidth set to 80kHz measured 0.00092%. Note: no significant 3rd harmonic observable. (c) As (b), but THD meter BW set to 20kHz; reading 0.00062%.*

device available. The Analog Devices AD797 quotes a THD of 0.0001% (−120dB) at 20kHz. But this is a bipolar input type, optimised for source impedances less than 1K, which is not convenient in the present application.) Next (with the wipers of R_{24} and R_{25} set to ground), R_{20} was adjusted for minimum distortion at 20Hz, this control proving the most critical at low frequencies. It compensates for the input offset in the OTA, centring the signal in the latter and thus minimising second harmonic distortion. It should not be left too far off centre, as there is then the possibility of control signal breakthrough appearing at the OTA output, leading *in extremis* to instability of the control loop on the 2–20kHz range. The next adjustment was of R_{31}, which as already stated was included as a way of out-phasing any third harmonic at the output. However, as noted in an earlier design brief[4] the main distortion mechanism in the op-amps used is second order, and in fact there was no visible third harmonic to outphase! R_{31} was therefore set to minimise the second harmonic content at 10kHz. This just leaves the circuitry associated with R_{24} and R_{25}, which has not been mentioned up till now. Op-amp A_{3a} provides a means of deliberately producing a soupçon of second harmonic distortion of any amplitude and at any relative phase angle over the full 360° of the fundamental. This is added in via R_{33} to the virtual earth at the input of A_{2b}. With suitable settings of R_{24} and R_{25}, the distortion at the main output at 600Hz can be driven down to the point where the residual reads the same as circuit noise (0.00016% with the 600Hz notch channel LPF in circuit, THD Meter BW 20kHz, OTA disabled so the oscillator is not oscillating), but this is an academic exercise. A better overall result was obtained by a compromise setting which resulted in somewhat more distortion at 600Hz but a substatial improvement (down from 0.0016%) at 10kHz. It might be thought that the presence of D_2 would make the performance unduly temperature sensitive, so a test was undertaken with the oscillator's top cover removed. The 600Hz distortion residual was monitored and a hair dryer played gently into the case, monitoring the temperature adjacent to D_2. At 50°C the distortion was unchanged from the room temperature value. The particular settings of R_{24} and R_{25} used resulted in 1.5V pk-pk at the output of A_{3a}, the waveform exhibiting a per cent or so of second harmonic distortion. This is diluted by a factor of about 300 by the ratio of R_{33} to R_{34} and a further factor set by the ratio of 1.5V pk-pk at A_{3a} output to the 10V pk-pk at LP, to around 0.0003%.

A number of practical details which arose may be of interest. An attempt to include its power supply within the case of the oscillator was a resounding failure, due to hum resulting from the mains transformer's stray magnetic field. So it was mounted instead in a 13A plug-style enclosure, with the low voltage secondaries piped to the rectifier, smoothing and stabilisers within the case of the oscillator.

Having produced a basic oscillator, it seemed sensible to include as many additional facilities as practicable. The first of these enables the unit to be used as a quadrature oscillator – very conveniently done with an SVF-based circuit. The HP output forms the $0°$ phase output, taken via a 560Ω resistor to a front panel BNC socket. The BPbar output, obtained from BP by inverting stage A_{6b}, provides an output lagging by $90°$ – note that although A_{1b} and A_{2a}, the two integrators within the loop, each provide $90°$ lag, the BP waveform actually *leads* that at HP, as they are *inverting* integrators. Another 560Ω resistor driven from LP provides the $180°$ output at a third front panel BNC socket. The instrument is also provided with an input, via a front panel socket and R_1. This may be used in two different ways. Firstly, with the oscillator running, its frequency can be locked to a low level signal injected via R_1, simply by tuning it to the frequency of the injected signal. Secondly, the oscillator can be disabled by setting S_2 to the 'Filter' position. This has the effect of spilling I_{abc} to the negative rail, disabling the positive feedback via the OTA and stopping the oscillation. At the same time, damping is applied from BPbar via R_{37} and R_{38}, defining the Q of the filter as $(R_{37} + R_{38})/R_3 = 30$. Since $R_1 = R_3$, the bandpass gain from IN to the $90°$ OUT socket (BPbar) should simply equal Q, the measured value at 600Hz actually being 27.

R_8 is a 2K wirewound two-gang potentiometer fitted with a ten-turn counting dial. As 10 on the dial corresponds to 200Hz, 2kHz or 20kHz it can act as a frequency readout, if the dial reading is simply doubled. A dial reading of 10 can be arranged to correspond exactly to 2kHz by fitting a select on test resistor between the top end of R_{8a} (and R_{8b}) and the op-amp output driving it. By selecting a different resistor for each frequency range (using additional poles on S_1), the top frequency on each range can be correct, without requiring exact close tolerance values for $C_5 - C_{12}$. Frequency readout will be almost linear, with a maximum error of -3% at 5 turns, due to the loading of R_{11} (R_{21}) on the wiper. This parabolic error can be reduced to a much smaller cubic error by connecting a 15K resistor from the top of R_{8a} to its wiper (and likewise R_{8b}). Some further development of the design is planned, notably affecting the fine output level control. Even using a high quality conductive plastic or CERMET type, R_{41} is a possible source of excess noise in the output. A better arrangement would be a switch providing 1dB steps, with a 0–1dB continuously variable control provided by a potentiomenter varying the reference voltage at the junction of R_{15}, R_{16}.

Box

Assuming E_1 stands for the rms value of the fundamental of a distorted sinewave, E_2 for that of the second harmonic component etc., then the total harmonic distortion is given by

$$\sqrt{\frac{(E_2^2 + E_3^2 + E_4^2 \; . \; . \; .)}{(E_1^2 + E_2^2 + E_3^2 + E_4^2 \; . \; . \; .)}} \qquad (1)$$

By setting a reference level with a flat frequency response, corresponding to the denominator of (1), and then measuring the relative level of the signal with the fundamental component E_1 notched out, this is exactly what a THD meter measures – except that in the measurements, any noise, hum or other signal present is also inevitably included. Ideally, we might prefer to measure

$$\sqrt{\frac{(E_2^2 + E_3^2 + E_4^2 \; . \; . \; .)}{E_1}} \qquad (2)$$

but if all components E_2, E_3, E_4, etc., are less than 10% of the amplitude of E1, then the difference due to each is less than 1%. In low distortion measurements, where the harmonics are all much less than 1%, then in practical terms (1) and (2) are identical.

References

1. Rosens, R. 'Phase-shifting oscillator', *Wireless World*, Feb. 1982 pp. 38–41.
2. Linsley Hood, J. L. 'Portable distortion monitor', *Wireless World*, July 1972 pp. 306–308.
3. Hickman, I. 'Design brief: low distortion audio frequency oscillators', *Electronics World and Wireless World*, April 1992 pp. 345, 346.
4. Hickman, I. 'Design brief: audio op amp with its head in the clouds?', *Electronics World and Wireless World*, July 1992 pp. 579, 580.

11 MOSFET stabilises Wien amplitude

Traditionally, Wien bridge-based audio oscillators employed an R53 thermistor for amplitude stabilisation. But this is expensive, and alternative schemes using diode stabilisation result in higher distortion. This circuit uses a MOSFET – much cheaper than the thermistor – but retains the latter's low distortion.

There are many cases where a spot frequency test oscillator is required: for instance, signal generators commonly incorporate 1kHz and 400Hz modulation oscillators. This circuit produces a low distortion 1kHz audio tone, using a low-priced MOSFET in place of the commonly employed (but expensive) R53 thermistor for amplitude stabilisation.

The circuit is a conventional Wien bridge oscillator, except that the positive feedback to the non-inverting input of the op-amp is tapped down the shunt resistor $(R_9 + R_{10})$ of the Wien network, the negative feedback to the inverting terminal being attenuated by a slightly larger margin. As the oscillation builds up, the output charges up C_1 via D_1 and R_2, partially turning on the FET, whose drain slope resistance in series with R_7 appears in parallel with R_{10}. This equalises the positive and negative loop gains, stabilising the output amplitude at (typically) 3V pk-pk. This arrangement keeps the AC voltage appearing across the FET small, hence minimising distortion.

Without the effect of the additional distortion reducing network (e.g. FET source grounded), the total harmonic distortion (THD) of the output is a modest 0.4% and since the FET is basically a square-law device, it consists almost entirely of second harmonic. By adding R_3 and modulating the source voltage with an appropriate fraction of the output via C_2 and R_4, the distortion can be greatly reduced. C_3, R_5 adjust the phase of this modulation for minimum distortion. The residual distortion is then mainly

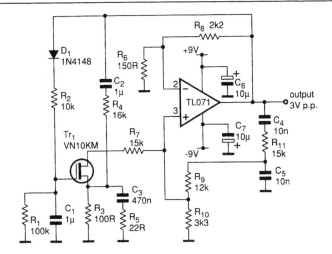

Figure 11.1 *Spot-frequency Wien-bridge oscillator uses a MOSFET instead of a thermistor for amplitude control, whilst retaining a THD of about 0.02%.*

third and higher harmonics, still due to the FET. A number of FETs were tested in this circuit, the recorded THD varying from just under 0.02% to a maximum of 0.025%. The output amplitude does not vary at all for supply voltage changes in the range ±6V to ±15V, though there is a slight rise in distortion relative to ±9V. Substituting 10K plus a 10K preset for R_4 will enable the circuit to be set up for minimum distortion at any supply voltage. The actual output frequency of the prototype circuit was 1009Hz. The resistors used should be 1% or 2% types and the capacitor tolerance should preferably be 5% or better.

12 High-performance THD meter

A THD (total harmonic distortion) meter, in conjunction with a very low distortion audio frequency oscillator, measures the distortion a signal suffers in going through a Hi-Fi amplifier or other audio equipment. It works by having 'zero' response (a very deep notch) at the fundamental of the test signal, but a flat response to all the harmonics. Using modern components, the design below provides a measurement range extending to well below 0.001%.

Serious work in any field of electronics requires the availability of test and measurement equipment adequate to the purpose. This is as true of the Hi-Fi field as any other, and an interest in this field has been a feature of this magazine since long before the days of the legendary Williamson amplifier. Even basic Hi-Fi testing, e.g. of an amplifier, requires a low distortion sinewave oscillator and a means of measuring the level of distortion in the amplifier's output. For the latter, a THD meter – total harmonic distortion meter – is very convenient, and has the advantage of being something the audio enthusiast can construct for himself for a modest outlay.

Figure 12.1(a) shows the schematic arrangement of a THD meter design dating from the early 1970s.[1] Figure 12.1(b) shows the full circuit, which is interesting not only because it shows what could be done with discretes in the days before the ready availability of op-amps, but also because it illustrates some of the problems in THD meter design. The complete instrument included a low-pass filter with choice of switchable cut-off frequencies to extend the lower limit of the measuring range by limiting the noise bandwidth, and provision for selecting a high-pass filter to reject hum (not shown). Styled a distortion monitor, it was intended for use with a separate external AF voltmeter, preferably true rms responding.

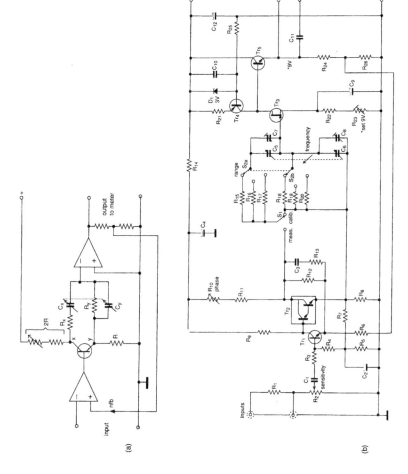

Figure 12.1 (a) Schematic of a notch circuit with NFB to provide a (nearly) flat response by one octave either side of the notch. (b) Part circuit of the distortion monitor of Ref. 1.

The circuit used a Wien bridge arrangemnent to provide a notch or transmission zero at the fundamental of the sinewave test signal. However, the notch produced by this type of circuit is cusp shaped. Not only is it very narrow at the null, but its return to full output at very low and very high frequencies is leisurely. Some quick sums show it to be still 7.5dB down at the second harmonic of the notch frequency. This is because of the low Q of the Wien network (the two Cs and two Rs, not the complete bridge), this being well below unity, in fact just 1/3. Consequently, it is necesary to include the notch circuit in an overall NFB (negative feedback) loop, to bring the response at the second harmonic to (ideally) much less than 1dB down. But the effective absence of NFB at the frequency of the test signal when the notch is correctly tuned to it, means that the fundamental is considerably accentuated in the stages within the loop. Thus it is necessary to keep the input amplitude well below the overload level of these stages to avoid the distortion meter introducing distortion of its own. A consequence is that the noise floor can limit attempts to measure very low THD levels. A further consequence of employing NFB is to narrow the notch, making tuning even more critical – the original article recommended a slow motion drive with at least a 100:1 reduction ratio.

These and other considerations limited the measurement range of the instrument of Ref. 1 to the order of 0.01% at 1kHz. I designed and built some years ago a THD meter with ranges down to 0.01% full scale, permitting readings down to around 0.002% or lower. It employs an SVF (state variable filter) based circuit which, having a higher Q, does not need so much overall NFB to obtain a reasonably flat response at the second harmonic. Even so, measurements at the limit of sensitivity are very tricky due to the narrowness of the notch, see Box. In consequence, when making the THD measurements on the low distortion oscillator described in Ref. 2, the THD meter was preceded by an auxiliary test circuit. This consisted of a fixed frequency twin-TEE network which was followed by a two-pole Chebychev high-pass filter. The tuning and peaking of the latter were adjusted to give a sensibly flat response, in conjunction with the notch, from the second harmonic upwards. Three such circuits permitted spot frequency testing at 20Hz, 600Hz and 10kHz.

Whilst this arrangement provided THD measurements which are reliably accurate, due to the suppression of the fundamental in a passive network before it meets any active circuitry, the restriction to spot frequency testing is in practice a serious drawback. So for many years I have been planning to replace my existing distortion meter with an improved design offering continuous tuning and a wider notch. The latter is needed since for measurements down towards the 0.001% level, adequate suppression of the fundamental is not possible with a single two-pole notch circuit. As indicated in the Box, to measure even 0.01% THD (requiring suppression of the fundamental to at least 0.003%) implies an accuracy of

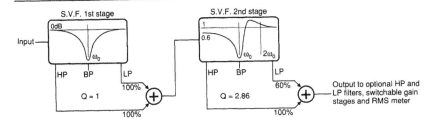

Figure 12.2 *Schematic arrangement of an improved front end of a THD meter, providing reduced internal distortion and less critical tuning of the notch.*

tuning of 15ppm, or 0.015Hz at 1kHz. This required accuracy is not an absolute figure, it is relative to the frequency of the sinewave oscillator providing the test signal. Even if the stability of the notch tuning were perfect, the oscillator may not exhibit the necessary long-term stability to permit readings to be taken. And even if it did, the short-term stability of an RC oscillator is likely to be inadequate. The inevitable close-in noise sidebands will not be adequately suppressed by the notch, or – if you prefer to think in the time domain – the frequency of the oscillator will shuffle about by a miniscule amount. This results in the fundamental peeping out randomly on either side of the notch, preventing a steady reading representing the harmonics only.

The solution presented here is to use two notches in tandem, greatly reducing the suppression required of each. This four-pole arrangement also permits a design avoiding the accentuation of the fundamental within the loop – necessary with a two-pole notch circuit to achieve a response at the second harmonic which is no more than, say, 1dB down on 'flat', i.e. on the response far from the notch. The scheme is shown in outline in Figure 12.2, where the first stage is an SVF notch circuit with a Q of unity. Thus there is no accentuation of the fundamental, the HP (high-pass), BP (band pass) and LP (low-pass) responses all being unity at the tuned (notch) frequency, see Figure 12.3. With the chosen Q of unity, there is a slight peak in the LP response, of just over 1dB at 62% of the tuned frequency (and at 1.6 times the tuned frequency in the case of the HP response), but in THD testing there is no signal present at this frequency. As the LP and HP outputs are in antiphase, summing them produces a notch at the tuned frequency.

The first stage sums equal contributions from the HP and LP outputs, resulting in a symmetrical notch, which with the chosen Q of unity is just 1.6dB down at twice the notch frequency. The second stage sums the HP output with just 60% of the LP output. This results in the notch occurring below the tuned frequency, with the low frequency response being only 60% of the response at high frequencies, as indicated in Figure 12.2. Furthermore, the notch now occurs at a frequency below the SVF stage's

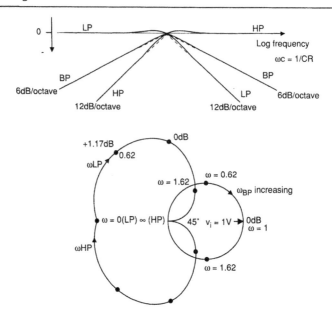

Figure 12.3 *(a) Frequency response of the LP, BP and HP outputs of a state variable filter with a Q of unity (Bode plot). (b) As (a) but shown as an Argand or vector diagram. (Reproduced from Analog Electronics, Ian Hickman, by courtesy of Butterworth-Heinemann.)*

resonant frequency. The resonant frequency is that at which the LP, BP and HP responses are all equal, and is given by $f = 1/(2\pi CR)$. By choosing a smaller value of CR for the second stage (in conjunction with the chosen ratio of LP to HP contribution), its notch can be arranged to coincide with that of the first.

By choosing a suitable value of Q for the second stage, a peak will occur at its resonant frequency, i.e. somewhere above the notch, and its amplitude can be made +1.6dB at twice the notch frequency, compensating for the −1.6dB response of the first stage. In fact, by judicious adjustment of the second-stage resonant frequency, ratio of LP to HP contribution and Q, the overall response can be made flat at the second and all higher harmonics of a sinewave test signal whose fundamental is suppressed by the two coincident notches. The arrangement has certain similarities to a four-pole elliptic high-pass filter, but with a major significant difference. Instead of spacing the two frequencies of zero response apart so as to maintain a designed stopband attenuation A_s all the way down to 0Hz, they are made coincident, since there is only one signal in the stopband, namely the fundamental of a sinewave test frequency. The harmonics all lie in the passband, which in the design presented here is flat to within less than 0.1dB.

The two stages were made up in temporary form, as per Figures 12.4(a) and (b), and tested. Note the tuning arrangement using a potentiometer to drive each integrator's input resistor R. This scheme using a fixed R provides linear tuning, avoiding the poor resolution at the high frequency end of the range which results if tuning is effected by varying R. Figure 12.5(a) shows the notch output of the Figure 12.4(a) circuit (set to 1kHz) at 10dB/vertical division, the span being 0–5kHz. Over 60dB of rejection was observed, the residual being due to the LP and HP outputs being not quite in antiphase: 60dB down corresponds to a departure from 180° of just 0.053°. Also shown is the LP output at 1dB/div., showing the expected peaking of about +1dB. Note that, for clarity, the traces are offset: the 0Hz levels of the two traces do not correspond. Figure 12.5(b) shows the same results, this time with the notch trace at 1dB/div. and the LP at 10dB/div. Compared to the response at 5kHz and higher, it can be seen that the response at 2kHz is indeed 1.6dB down as predicted by theory.

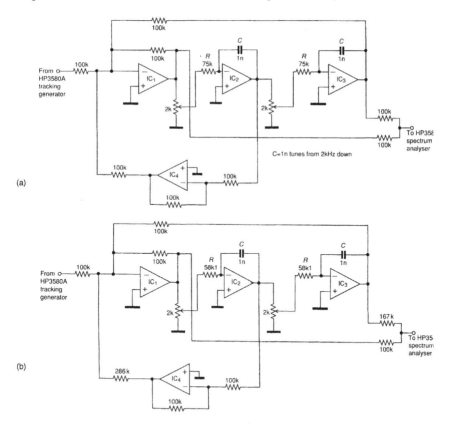

Figure 12.4 *(a) Circuit of the first stage of the improved THD meter. (b) Circuit of the second stage of the improved THD meter.*

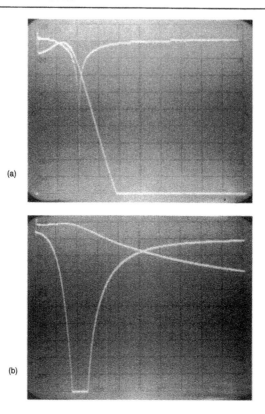

(a)

(b)

Figure 12.5   *(a) First-stage response; span 0–5kHz, notch trace 10dB/div., LP trace 1dB/div. (b) As (a), but notch trace 1dB/div. and LP trace 10dB/div.*

Figure 12.6(a) shows the response of the second SVF stage, with its tuning set so that the notch again occurs at 1kHz. The stage's resonant frequency is actually 1.29kHz, with around 10dB of peaking at the LP output. This results in a smaller peak in the notch output, the response still being +1.6dB at 2kHz relative to the far-out high frequency response. When this is combined with the first-stage response of Figure 12.5, the result is a response which is level at 2kHz and upwards, see Figure 12.6(b). The notch should be 120dB or more deep, but as displayed it is limited to the spectrum analyser's noise floor at −90dBref. To reliably achieve 60dB or more suppression in each stage, putting the fundamental of the test sinewave below noise, a phase trim should be provided for each stage, similar to that shown for the second stage, in Figure 12.7.

A complete THD meter front end must include facilities to accept inputs of various amplitudes, so some kind of input attenuator is required as indicated in Figure 12.7. The potentiometer used should be a wirewound type, to avoid introducing noise. The twin-gang tuning potentiometers

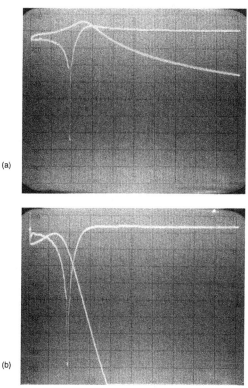

(a)

(b)

Figure 12.6 *(a) Second-stage response, notch trace and LP trace 10dB/div. (b) Combined four-pole notch response 10dB/div. at output of second stage, with first-stage LP response at 1dB/div.*

should also be wirewound, for the same reason. Also shown is a 'SET LEVEL' switch to permit the meter deflection to be set to full scale on the incoming signal, before notching out the fundamental. Figure 12.7 shows a fine-tuning trim on the second SVF stage. This is optional but assists in obtaining the maximum possible fundamental rejection. It is necessary if the twin-gang tuning potentiometers for the two stages are ganged together. In this case, an 'INITIAL TUNE' position is needed, permitting tuning of the first stage for maximum rejection. The second stage is now also (approximately) tuned, due to the ganging, and final adjustment of the second-stage frequency and phase trims completes the tuning. Dual-gang 2K wirewound pots are available to special order* with the shaft extended at the rear, permitting the tuning of the two SVF stages to be ganged.

Due to the 20dB gain stage between the first and second SVF stages, the lower limit of the measurement range is set only by first-stage noise. In the

* Available from Spectrol Reliance Ltd, Tel. 01793 521351, Fax 01793 539255.

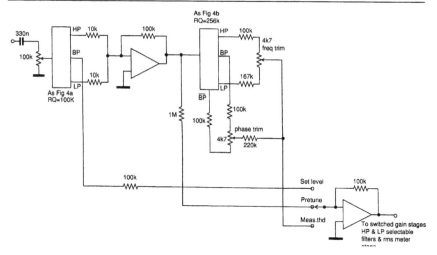

Figure 12.7   *Suggested complete front end for an improved THD meter.*

unlikely event that measurements of THD exceeding 10% may be required, provision must be made to switch the 20dB gain stage to 0dB.

The results shown in Figures 5 and 6 were taken using TL084 quad
op-amps in circuits put together on Experimentor® plug boards. For the
purpose of proving the principle, these were more than adequate. However, for a finished instrument these op-amps would be unsuitable, due to
their THD of around 0.003% typical. A better choice is the Burr-Brown
OPA2604 dual FET-input audio op-amp, with its 0.0003% THD typical.
Clearly, careful construction and screening between stages will be necessary to achieve 120dB or more of fundamental suppression. Given this, the
limiting factor on readings is likely to be noise and hum. The former can be
reduced by a low-pass filter immediately preceding the measuring circuit,
with switchable cut-off frequencies of, say, 200, 80 and 20kHz, the latter by
switching in a high-pass filter with heavy attenuation at 50Hz.

Box

Figure 12.8 shows the circuit of the classic four op-amp state variable
filter (SVF). There is a three op-amp variant, but this involves taking the
damping term from the bandpass output back to the non-inverting
input of IC_1, resulting in a common mode component at the filter's
input. The tuned frequency or frequency of maximum gain at the
bandpass output is given by $f = 1/(2\pi CR)$. At this frequency the HP, BP
and LP outputs are all equal in amplitude, with the BP and LP outputs
leading the HP output by 90 and 180° respectively.

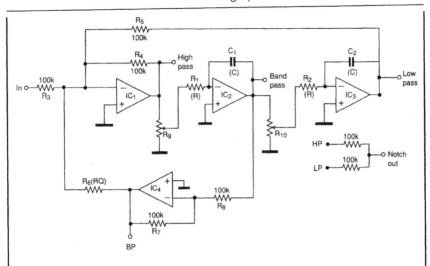

Figure 12.8 *Classic four op-amp state variable filter, with improved tuning arrangement.*

With the resistor values shown, the low frequency gain at the LP output and the high frequency gain at the HP output are both unity, whilst the gain at the tuned frequency is numerically equal to the circuit Q, where $Q = RQ/100k$. The transfer function is given by numerator/ $(s^2 + Ds + 1)$, where the numerator equals 1 for the LP output, s for the BP and s^2 for the HP output, and $D = 1/Q$. Now s is the complex frequency variable $\sigma + j\omega$, but for the purposes of determining the steady state response of the circuit to sinewaves, σ can be ignored. This leaves just $j\omega$ as the variable, where $\omega = 2\pi f$, ω being in radians per second.

Things can be simplified even further by normalising the frequency, that is, simply assuming that whatever tuned frequency one is interested in is unity. Thus the gain at the low-pass output is given by $1/(-\omega^2 + Dj\omega + 1)$. At the tuned frequency, where $\omega = 1$, this amounts to $1/jD = -j(1/D)$ where the $-j$ indicates a phase lagging 90° on the input. So at the tuned frequency, if $D = 1/2$ (i.e. $Q = 2$), the gain is ×2 or +6dB at −90° relative to the input. (This is the case for a theoretical second-order response. But the two integrators in the SVF are actually inverting integrators, so in this circuit, the bandpass output in fact leads rather than lags the high-pass output.)

The notch output of the circuit of Figure 12.4(a) is obtained by summing the HP and LP outputs, so the gain is given by $(-\omega^2 + 1)/(-\omega^2 + Dj\omega + 1)$ – clearly the numerator is zero at $\omega = 1$. On substituting $D = 1$, $\omega = 2$, the gain turns out to be ×0.83 or −1.6dB, this being the response of the circuit at the second harmonic of a sinewave test signal, when the notch is tuned to the fundamental. In the Figure 12.4(b) circuit, D is set to 0.35 (Q = 2.86) and the contribution from the LP output is reduced to 60%, so the gain is given by $(-\omega^2 + 0.6)/(-\omega^2 + 0.35j\omega + 1)$, giving the response illustrated in Figure 12.6(a). Note that in this case, though, $\omega = 1$ corresponds to 1.29 times the $\omega = 1$ of the first SVF section, i.e. to 1.29kHz for a 1kHz test signal, making the notches of the two stages coincident.

The sharpness of the notch in the circuit of Figure 12.4(a) can be found by a little judicious approximation of the expression $(-\omega^2 + 1)/(-\omega^2 + j\omega + 1)$ – remember D = 1 for this circuit. At $\omega = 1$, the numerator is zero and the denominator is jω, or just unity (amplitude-wise). If the frequency is changed by 0.1%, the denominator is virtually unaffected. However, with ω now equal to 0.999, the numerator becomes 0.002, or only 54dB down on the response at $\omega = 0$ or infinity. Thus for a fractional detuning from the notch of δ, the output rises from zero to 2δ. For reasonably accurate THD measurements even down to a modest 0.01%, the fundamental must be suppressed to 0.003%. So the accuracy of tuning must be at least 0.0015%, or ±0.015Hz at 1kHz.

References

1. Linsley Hood, J. L. 'Portable distortion monitor', *Wireless World*, July 1972 pp. 306–308.
2. Hickman, I. 'Low distortion audio oscillator', *Electronics World and Wireless World*, May 1994 pp. 370–376.

13 Listening for clues

A 'squawk-box' is very handy around the lab., and a rather more 'Hi-Fi'
version even more so. This version incorporates input protection, an AF
millivoltmeter, and other features, making it still more useful.

Introduction

Every electronics laboratory in which I have ever worked has had a
general-purpose laboratory amplifier or 'labamp' – invaluable for hooking
on to a circuit (be it a detector, discrimintor, DAC output or whatever) to
hear what is going on. One might think that the more obvious option is to
see what is going on, with the aid of an oscilloscope. But if there is a mixture
of noise, hum and possibly other signals as well, it may be difficult or
impossible to interpret the display, or even to trigger the oscilloscope so as
to obtain a coherent picture. With a mixture of signals present, an audio
frequency spectrum analyser might be more appropriate, but few labs
possess such an instrument. But wait a minute, most electronic engineers
possess not one but two AF spectrum analysers (not supplied by the
company), one situated on each side of the head. Hence the utility of the
universal labamp.

Most of the labamps one comes across have been knocked up in a hurry
by an engineer, when the need arose, and they typically consist of an
amplifier and a small loudspeaker, housed in a die-cast box and powered
by an internal dry battery. With its tinny low-fidelity reproduction, such a
device is generally known as a 'squawk-box', and very useful it can be, too.
In fact, its advantages are as numerous as its disadvantages. For example,
running from an internal battery means that it is immune to the hum
problems which might, in a mains-powered version, be caused by earth
loops. But unfortunately, just when you need it in a hurry it usually
transpires that the batteries are flat, because the last user left it switched on.
Then again, it would be useful to be able to hear whether the signals in the
circuit under observation are corrupted by hum (*not* caused by an earth

loop), but with its small loudspeaker, the typical squawk-box remains silent upon this topic.

A new approach

Some while ago, I resolved to replace my then squawk-box (the last in a long line, mostly converted from superannuated radios) with a version having a decent frequency response and a generous output of a few watts. This indicated mains operation, but with precautions to avoid the possibility of hum due to earth loops. And whilst building an audio amplifier system, a calibrated input step attenuator and meter circuit could be incorporated, making the unit double as an AF millivoltmeter. In addition, a 600Ω unbalanced signal output would be provided, permitting use as a hum-loop-free general-purpose preamplifier. For good measure, access would be provided to the loudspeaker's voice coil, to permit the unit to be used also as an extension speaker, whilst as an alternative, the amplifier's output would be available to drive an external speaker.

The input stage

Whilst the unit would be provided only with an unbalanced high impedance input, the actual input stage would be balanced, permitting the rejection of any hum present on the 'earthy' input-low line, on which the wanted signal would be riding. The input stage therefore used a conventional three op-amp instrumentation amplifier configuration, as shown in Figure 13.1, using three quarters of a TL081 quad op-amp. The gain of the input pair to balanced or 'push-pull' signals is equal to $(2R_1 + R_g)/R_g$, whilst their gain to common-mode or 'push-push' signals is unity, i.e. these appear unaltered at the output of the input pair. Thus whilst the input pair provides no common-mode rejection as such, the balanced-to-unbalanced signal ratio is improved by the ratio $(2R_1 + R_g)/R_g : 1$, which could be large. The output of the input pair is applied to the input of the third amplifier, whose gain to balanced signals is R_3/R_2. However, assuming the two R_2s and two R_3s are exactly matched, they form a bridge circuit, so that the common-mode component appearing at the inverting input of the third op-amp exactly equals that appearing at the non-inverting input. Thus the overall common mode rejection ratio CMRR is that provided by the third op-amp, times the addtional ratio $(2R_1 + R_g)/R_g : 1$ mentioned earlier.

In the present application, the wanted signal appears between the input terminals in unbalanced form, but the circuit still responds to the *difference*

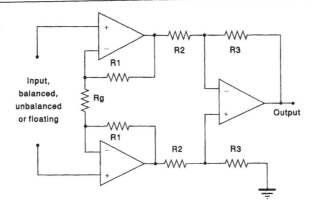

Figure 13.1 *Basic three op-amp instrumentation amlifier. Overall effective CMRR depends on the ratio of R_1 to R_g, amongst other things.*

voltage between the two terminals. However, any hum (due to an earth loop) on the input-low terminal (the outer of the BNC input socket) will appear also on the input-high lead or centre pin of the socket. It is thus a common-mode component and will be rejected as described above.

The AF millivoltmeter stage

The AF millivoltmeter stage which was included in the instrument is a simple one, using a full-wave rectifier circuit, and scaled to read rms when the input is a sinewave. Obtaining a linear scale was at one time difficult, due to the forward volt drop of the necessary diodes. Various schemes were formerly used, from individually calibrating the meter scale to allow for the diode non-linearity, to using a high impedance (such as the collector output of a transistor) to approximate a constant current source. The circuit used here appeared in *Wireless World* many years ago, and encloses the meter and a bridge rectifier in the feedback loop of an op-amp, Figure 13.2. Assuming the open loop gain of the op-amp remains high up to the highest frequency of interest (20kHz, in this case), it will force the voltage at the inverting terminal of the op-amp to follow that at the non-inverting input, in the process forcing a current defined by the lower resistor through the meter, regardless of the volt drop across the diodes (which will in any case vary slightly with temperature). As the input voltage passes through zero, the op-amp becomes momentarily open loop. Just a small voltage difference between its input terminals forces it to slew as rapidly as it is able until the other side of the diode bridge turns on, restoring closed loop operation.

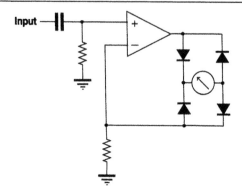

Figure 13.2   *Simple AF millivoltmeter with linear meter scale.*

The complete instrument

This is shown in Figures 13.3 and 13.4. Figure 13.3 shows the input at a BNC socket applied via a DC blocking capacitor and a 4K7 safety resistor to a range switch S_2. This, in conjunction with the gain of the following stages, provides nine input ranges giving FSD (full scale deflection) factors for the millivoltmeter function of 3mV FSD to 30V FSD. The input of the op-amp connected to the wiper of S_2 is protected by two back-to-back diodes. These are rated at 75mA peak current, which corresponds to a peak input voltage of about 350V. But as the peak dissipation in the 4K7 safety resistor under these circumstances would be over 25W, this should be regarded as only a momentary withstand voltage, or a 4K7 resistor of the fusible variety could be used.

The input appears between the non-inverting inputs of the first stage of the instrumentation amplifier, which provides a gain of ×20. When monitoring an earth-free source, e.g. a piece of battery-operated kit, S_1 can be closed, providing an earth for the item under test. Where a hum-loop problem is encountered, S_1 should be opened, breaking the loop. The associated 15K resistor provides a 'static drain', to keep the input amplifier earth-referenced, even if the input socket is left open-circuit. The associated zeners limit any ground-line float, when connected to other equipment, to just over ±3V pk-pk – if the difference between grounds on the monitor and any equipment to which it is connected is greater than this, it is suggested that further investigation is advisable, on possible safety grounds.

The third section of the quad op-amp provides a gain of around ×6, the 2K2 potentiometer permitting an adjustment for maximum common-mode rejection. This is obtained when the ratio of input to feedback resistor on the inverting side equals the potentiometer ratio on the non-

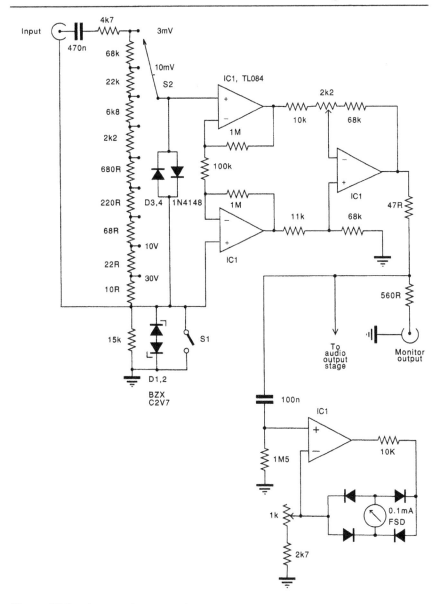

Figure 13.3 *A general purpose laboratory amplifier/AF millivoltmeter: input and millivoltmeter stages.*

inverting side. The output of the instrumentation amplifier stage is made available at a BNC output socket, labelled 'Monitor', at an impedance of 600Ω unbalanced. The outer of this socket is connected to the circuit's 0V line, and hence is referenced to the input socket outer, or the instrument's

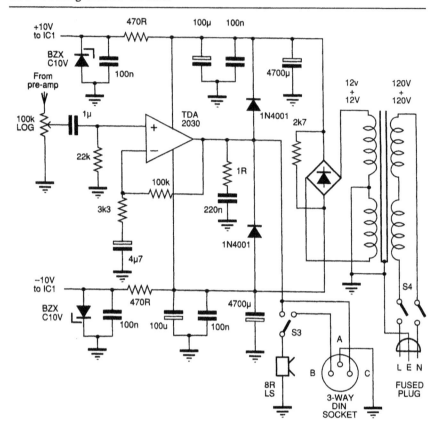

Figure 13.4   *A general-purpose laboratory amplifier/AF millivoltmeter: power amplifier and power supply stages.*

mains supply earth, according to whether S_1 is closed or open. The low frequency −3dB point of the input amplifier is 3.2Hz, much lower than is necessary for the reproduction of which the loudspeaker and enclosure is capable, but it was chosen to provide a wider-than-audio frequency response at the monitor output.

The third section of the quad op-amp also drives the meter stage, which uses the final section of IC_1. Here again, the 100n coupling capacitor and 1M5 resistor provide a frequency response extending below the bottom of the audio range. Germanium gold-bonded diodes were used in the meter circuit, in the interest of low forward volt drop. The overall sensitivity is set up with the 1K pot. The 10K resistor driving the bridge plays no useful part in normal operation, but limits the current applied to the meter when a large input overload is applied. Whilst the inclusion of this resistor is not good for the frequency response, the instrument is nevertheless flat from 20Hz to 10kHz, and less than 1dB down at 20kHz.

Figure 13.4 shows the output amplifier and power supply stages. The power amplifier used is the TDA2030, which, for convenience, was mounted with its associated components on the matching ready-made PCB, RS434-576. The power supplies, input stage and millivoltmeter stage, on the other hand, were all constructed on a piece of 0.1 inch matrix copperstrip board.

The loudspeaker volume is controlled both by the setting of the input attenuator S_2 and by the 100K log pot volume control at the input of the TDA2030. (The ON/OFF switch S_4 is ganged with the volume control.) The TDA2030 amplifier drives a wide range twin-cone loudspeaker type RS249-031, mounted in its matching cabinet RS249-801. This loudspeaker may alternatively be switched to a 3-pin socket, via which it may be used as an extension speaker, or as one of a stereo pair with a Hi-Fi. For further versatility, the amplifier may be used to drive an external loudspeaker connected to pins A and C of the three-way socket, either by itself or in parallel with the internal speaker, according to the setting of S_3.

The two-rail power supply is entirely conventional, providing about ±17V to the TDA2030 under quiescent conditions. This is dropped to a stabilised ±10V for the preamplifier and millivoltmeter stages. A load resistor is fitted to ensure rapid discharge of the smoothing capacitors on switch-off, preventing a possible nasty surprise if the unit is opened up and worked on for servicing.

Practical considerations

With its fully enclosed cabinet, the loudspeaker creates quite some pressure inside the enclosure when reproducing low frequencies at volume. Care is therefore needed with construction if rattles are to be avoided. The two BNC sockets, the meter and S_1 were mounted on a small Formica panel covering the lower third of the grille cloth, clear of the loudspeaker cut-out. The panel was firmly screwed into place, with the rear of the components projecting back through the front panel of the cabinet, in holes just large enough to accommodate them. S_1 and the volume control were mounted on an aluminium subpanel the same size as the Formica panel, but mounted behind the enclosure's front panel, which had holes just large enough to accommodate the shafts of the controls, rebated on the inside to clear the nuts and bosses. The aluminium panel also had a hole to clear the rear of the meter, and small holes to pass the leads to/from the BNC sockets and S_1.

Some care is needed to avoid hum pickup, especially on the more sensitive ranges. So inside the cabinet, aluminium foil was fitted to cover half of the bottom of the case and the whole of the side where the stripboard input amplifier/AF millivoltmeter was mounted. Like the aluminium subpanel, this foil was connected to the power supply 0V rail. The mains transformer was bolted firmly to the base of the cabinet, on the opposite side from the input board.

Conclusion

The labamp described has been in use for over ten years now and has proved entirely reliable. It has also been found extremely useful in the laboratory, in addition to its more usual everyday use as the right-hand speaker in a modest 'Hi-Fi' system. With its wide frequency response, it is much more informative than a small diecast-box-housed battery-operated squawk-box when used to monitor activity in an audio circuit. A simple diode probe has also been made, permitting signal tracing in RF and IF circuits.

Part 3
RF

14 White noise – white knight?

This article describes a simple, inexpensive calibrated RF noise generator for use in the frequency range 1–1000MHz. Such an instrument, in conjuction with a 50Ω variable attenuator, permits the rapid and accurate measurement of receiver noise figures. (Regular readers of *Electronics World and Wireless World* may recall the occasional appearance of an electronic chat column called 'White Noise', by 'Hot Carrier', and may perhaps guess at the identity of the latter.)

No, this is nothing to do with the ramblings of Hot Carrier, but I make no apology for visiting again a topic which was covered in an earlier Design Brief.[1] Noise is an unwanted guest in all our analog circuits, so it seems perverse to want to make yet more noise; but the reason is simply summed up in the old adage 'Know thine enemy'. A calibrated noise source is a great convenience when developing a low noise circuit, and – although other methods do exist – is almost essential for the accurate determination of the noise figure of a low noise amplifier or receiver. The classical approach to the design of a wideband noise generator is to employ a 'temperature limited' thermionic diode, that is to say one where all the electrons emitted from the cathode are attracted directly to the anode: there is no cloud of electrons forming a 'space charge' surrounding the cathode. This is usually achieved by using a pure tungsten filament as the cathode, fed with a smooth DC current by the adjustment of which the anode current may be set to the desired value.

The resultant noise is described as 'shot noise' since the electrons forming the anode current fall onto the anode like lead shot onto a corrugated iron roof. If furthermore there is no residual gas in the valve which could be ionised and the anode voltage is not so high that secondary emission results, the rms fluctuation (noise) current i is related to the DC anode current I, according to the formula

$$i^2 = 3.18 \times 10^{-19} \times I \times df$$

where I is in amperes and df is the bandwidth of interest. By contrast, the noise current i flowing in a short-circuit across a resistor R, is given by

$$i^2 = (1.59 \times 10^{-20} \times df)/R$$

so that a noise diode passing an anode current of I gives as much rms noise current as an equivalent resistor $R_{eq} = 0.05/I\Omega$. The noise current available from a noise diode permits the construction of a noise source whose output forms a known, absolute standard. Figure 14.1(a) shows the arrangement, where the anode current is passed through a 50Ω resistor, thus forming a source of noise matched to the circuit under test, the magnitude of the noise power relative to that of a 50Ω resistor being precisely known. In theory, df can extend from 0Hz to as high a frequency as one wishes,

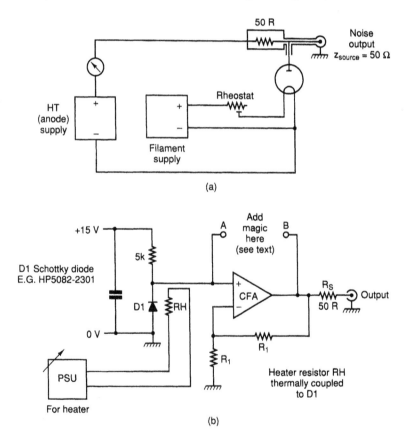

Figure 14.1 *(a) Simplified circuit diagram of a wideband RF noise generator employing a temperature-limited diode. (b) Suggested RF noise generator using a Schottky diode.*

although in practice an upper limit is set by the shunt capacitance to ground of the anode of the noise diode and other factors. Rhode and Schwarz are probably the best-known European manufacturer of noise sources of this kind, the model number (from memory) being SKTU and the specified operating range up to 1000MHz. The measurement of receiver noise figure is simplicity itself. With the noise generator connected to the receiver under test and the diode anode current zero, the noise output from the receiver is noted, using a suitable measuring instrument. 3dB of attenuation is then inserted between the receiver output and the measuring instrument, and the output from the noise generator increased to restore the indication to that previously noted. The noise supplied by the noise generator is then equal to the receiver's own front end noise. A meter in the noise generator monitors the anode current and is directly calibrated in dB above thermal, corresponding values of kTB also being given. Practical considerations in the manufacture of the noise diode limit the maximum output that can be achieved to around 15dB above thermal, so that if it is desired to measure the noise figure of a receiver which is higher than this, the output of the noise generator must be amplified first. Clearly the amplifier used for this purpose will need to have a good signal-to-noise ratio, or since in this instance the 'signal' is itself noise, a good noise-to-noise ratio!

A solid state alternative to the thermionic diode noise generator was described in these pages by T H O'Dell.[2] Here, the source of noise was the flow of reverse 'leakage' current induced by the creation of hole/electron pairs in a photodiode, by photons illuminating the same. It was stated that the reverse photodiode current is subject to the same relation between DC and its shot noise component as the saturated thermionic diode. However, as the practical limit to the available current, at 100μA, was only about 1% of the maximum current possible with the thermionic diode, it was necessary to work at an impedance level of around 5K and provide a 10:1 turns ratio transformation to 50Ω. As just 1pF of stray capacitance across a 5K source would give a −3dB point of 32MHz, a tuned transformer was used, this having the additional advantage of absorbing the self-capacitance of the diode as well as that of the transformer. Not having a sample of the HP5082-4220 photodiode to play with, I wondered whether an HP5082-2301 (1N5165) Schottky diode could be used. The paint was scraped off and the diode's reverse leakage current at $V_r = 15V$ found to be totally unaffected by any practical level of illumination. However, there is another mechanism for inducing hole/electron pairs in a semiconductor – heat. The leakage of the sample diode at $V_r = 15V$ and (almost) 25°C was 0.056μA, against the manufacturer's quoted maximum of 0.300μA. Raising the temperature of the device by holding it close to, but not touching, a soldering iron, produced a leakage current of over 100μA, at an estimated temperature of between 200 and 250°C. This is beyond the device's top rated temperature of 150°C, but on being left to cool, the current fell to less than a microamp in seconds and right back to 0.056μA within a few minutes. Thus this scheme might form the basis of a practical noise generator, the utility of which would clearly be much increased if it were broadband, rather than limited to a spot frequency as in the design in Ref. 2. Figure 14.1(b) shows such a possible scheme. In addition to the diode's self-capacitance (of under 0.3pF at 15V reverse bias), there is the input capacitance of the amplifier to consider, typically several pF, which would limit the useful top frequency to less than 10MHz. However, given a modern low noise current feedback op-amp with a bandwidth at a non-inverting gain of two approaching 1GHz, the high frequency attenuation of the diode noise due to the total shunt capacitance could be avoided by cancelling it with an equal shunt negative capacitance to ground, as described in Ref. 3. Here, it just involves connecting the equal capacitance between points A and B in Figure 14.1(b).

Noise measurements are also necessary in audio circuits. Figure 1 of Ref. 1 shows three different audio frequency noise generators, while pointing out that they can conveniently be replaced by a simple digital noise generator, the National Semiconductors MM5437. This is featured in Figure 3 of the said article, which also showed the typical output noise in the time domain, i.e. waveforms. Figure 14.2(a) shows the output of the MM5437 in the frequency domain, indicating that the noise level is virtually flat to 60kHz and only 3dB down at about 100kHz. Note, however, that this is with no low-pass filtering of the pseudo random digital output from the chip, so that the amplitude distribution in not Gaussian. The omission of any low-pass filtering also makes the first zero of the output at 180kHz very obvious, with a spectral line due to clock break-through sitting in its middle. The spectrum repeats, at diminishing ampli-tude, around multiples of twice the clock frequency, so the device can be used as a noise source, over limited bandwidths, at frequencies well above that of the clock, up to low radio frequencies see Figure 14.2(b).

For higher radio frequencies, whilst a noise source based on a tempera-ture-limited noise diode source is convenient for lab measurements, it is over-large and expensive for building into an equipment. But one based on the shot noise in an illuminated photodiode or a heated Schottky diode, it has been shown above, is likely to be limited to operation at frequencies in the lower VHF range. A number of manufacturers specialise in producing compact semiconductor noise sources for OEMs to build into equipments, probably the best-known maker being an American company,[4] Figure 1(b) of Ref. 1 showed a zener diode used as a semiconductor noise source, with the comment that this can work up to VHF, given a suitable diode. There was a CV specification – I forget the number – for a zener diode selected for use as a noise generator. The specification only covered performance up to 100kHz, though the actual performance of the diode extended well

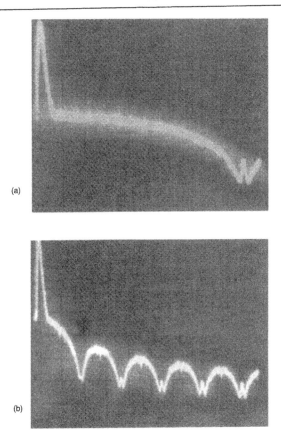

(a)

(b)

Figure 14.2 *(a) Output spectrum of National Semiconductors MM5437 digital noise generator (output 2, pin 5), top-of-screen reference level –20dBm, resolution bandwidth 3kHz, video filter max, 10dB/div vertical, 20kHz/div. horizontal, centre frequency 100kHz. The output of the MM5437 was not low-pass filtered, so some 180kHz clock breakthrough is visible in the first zero of the spectrum. (b) As (a) except resolution bandwidth 10kHz, 100kHz/div. horizontal, centre frequency 500kHz.*

beyond this. In the 1960s, it occurred to me that if one wanted perform-ance up to VHF, or better still UHF, it would make sense to start with a device specified for use at that sort of frequency. It is never a good idea to rely on characteristics of a device which are not specified, or even monitored on a sample basis, by the manufacturer: he is quite at liberty to introduce a process change that improves a zener diode's performance for its primary purpose but ruins it as a noise diode. So I experimented with what was at the time about the best UHF transistor available, a 2N918, running the base emitter diode in reverse breakdown with the collector left disconnected. With some qualifications, it worked well and in the 1970s I had it designed in as a noise diode in the BITE (built-in test equipment) of a military tactical radio system, as a GO-NO-GO test of the front end and IF modules. Figure 14.3(a) shows a circuit using a 2N918 thrown together to reproduce those results, the performance being as in Figure 14.4(a). The lower trace is with the diode current at zero, i.e. it shows the noise floor of the spectrum analyser, with a local TV station just making itself visible at around 510MHz. The top-of-screen reference level is −33dBm so the trace is at −93dBm, and (assuming the noise bandwidth of the analyser's 1MHz IF resolution bandwidth filter is equivalent to a 1MHz wide brickwall filter) this corresponds to −153dBm/Hz. Thermal being −174dBm/Hz, this equates to a noise figure of 21dB – rising to 22dB at 1GHz. This is good for a spectrum analyser, since these are always designed primarily for linearity rather than sensitivity; 25dB is a not untypical noise figure. With a diode current of 156µA, the power delivered to the analyser was about 12dB above the noise level, or about 33dB above thermal – a lot more than one can get from a noise generator using a thermionic noise diode. If the current was increased to 310µA, the noise level above 300MHz was

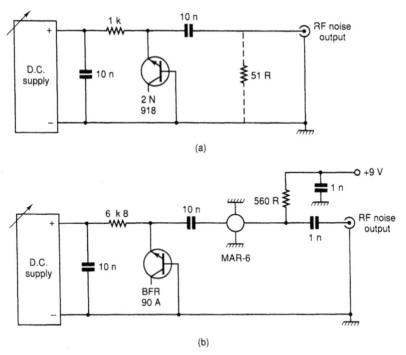

(a)

(b)

Figure 14.3 *(a) Circuit used to test the base-emitter junction of a 2N918 as an RF noise source. (b) Circuit of a noise source using the base-emitter junction of a BFR90A in reverse breakdown.*

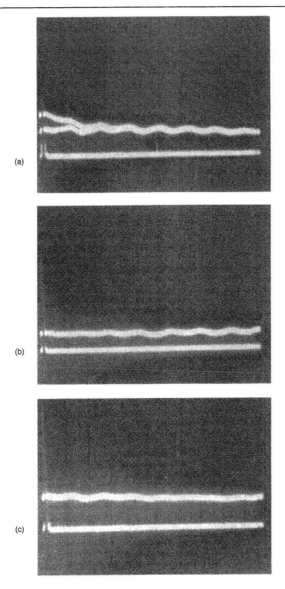

Figure 14.4 *(a) Output of the noise source of Figure 14.3(a): lower trace, diode current zero; middle trace, diode current 156µA; upper trace, diode current 310µA. (10dB/div. vertical, ref. level –33dBm, centre frequency 500MHz, span 1000MHz, IF (resolution) bandwidth 1MHz, video filter max.) (b) As (a), but a 51Ω load resistor added to the source. (c) Output of the noise source of Figure 14.2(b): lower trace, diode current zero; upper trace, diode current 1.3mA.(10dB/div.vertical ref. level –33dBm, centre frequency 500MHz, span 1000MHz, IF (resolution) bandwidth 1MHz, video filter max.).*

unchanged, but the level rose below that frequency. Further increases of current saw the picture alternating smoothly between the two upper traces. The noise was definitely falling off by 1500MHz, so the total noise delivered to the 50Ω load presented by the spectrum analyser's input was roughly −60dBm or a bit less, or around 0.001μW. With the base emitter breakdown voltage of 4.5V, 156μA represents a power input to the diode of 700μW, giving an efficiency as a noise generator of not very much, but still much higher than a thermionic noise diode. The variations of around ±1dB in noise output were a mystery at this stage. Their periodicity is around 150MHz, corresponding to a round journey in coax of around 65cm, but the circuit was connected to the analyser by a lead of no more than 6cm, including the BNC plug.

Figure 14.4(b) shows the effect of connecting a 51Ω resistor in parallel with the noise output; the variations in level have been largely damped out, especially up to 300/400MHz. The level has apparently fallen by 4 or 5dB, but is now so close to the analyser noise that the latter is contributing to the indicated level. So the true fall is probably nearer 6dB, which is what one would expect if the diode acted as a perfect constant noise-current generator. So we now have a matched 50Ω noise source, albeit at lower power. Where the noise power was $i^2 \times 50$, with the extra resistor in parallel it becomes $i^2 \times 25$, the rms noise current i being constant. Furthermore, not only is noise power halved, but half of what there is, is now dissipated in the additional 50Ω resistor, hence the 6dB drop in output. Another 2N918 was tried in this circuit with generally similar results, except that there was no value of operating current which would avoid some rise in output below 300MHz. As the 2N918 is a very ancient device, an obvious next step was to try a more modern transistor. A BFR90A was therefore connected in circuit, but found to give a much lower output than the 2N918, barely above the analyser's input noise. A 20dB broadband amplifier stage, using a Mini Circuits[5] Mar 6 type amplifier was therefore added, as in Figure 14.3(b), the resulting output then being about 13dB above analyser noise, see Figure 14.4(c). The variations of output level with frequency are now quite low right up to 1000MHz. This is not due to the different device but to certain other circuit changes. The 1k resistor in Figure 14.3(a) was an 1/8W miniature carbon film axial lead type. Such resistors are manufactured with a film of about 1% of the nominal value, required final value being achieved by making a spiral cut in the film. Such resistors thus have an appreciable inductance, though it is often possible to ignore this due to its low Q. The output shown in Figure 14.4(c) was achieved with a different feed resistor, Figure 14.3(b), namely a 6K8 sub-miniature solid carbon type, with a further substantial improvement by selecting the optimum supply voltage to the amplifier. With the aid of a hairdryer, the whole noise generator circuit was raised to +75°C, with no measurable change in output noise at any part of the 0–1000MHz range.

Whilst the circuit of Figure 14.3(b) offers the basis of a potentially useful noise source, its output level is fixed, not readily adjustable from thermal upwards as in the case of a thermionic diode noise source. However, this limitation is easily circumvented by the addition of a step attenuator. As the attenuation is increased from zero, the noise delivered to the circuit under test is reduced, in principle only reaching thermal when the added attenuation is infinite. In practice, as Figure 14.5 shows, 31dB is sufficient to reduce a noise level of 25dB above thermal to a mere 1dB above thermal, low enough to test any amplifier or receiver operating at room temperature, with the possible exception of a parametric amplifier. A noise output of 25dB above thermal was mooted, as it is sufficient for most applications (and certainly much more than obtainable from most thermionic diode noise sources), but well below the level available from the circuit of Figure 14.3(b). This allows for the fitting of a fixed 9dB pad at the output of the Mar 6 amplifier, which would be enough to provide a source with a return loss of 18dB even if the output VSWR of the amplifier were infinity. In fact, the output VSWR of the Mar 6 is 1.8 : 1 maximum up to 2GHz, corresponding to a return loss of 11dB, so the addition of a 9dB pad at the output would provide a noise source with an output VSWR of less than 1.08:1 – a return loss of around 30dB.

From the foregoing, it seems clear that a very economical semiconductor noise source operating up to 1000MHz, with a calibrated adjustable output flat to within ±1dB and suitable for the lab measurement of receiver

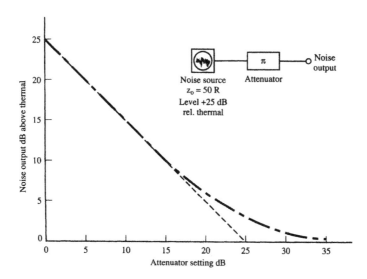

Figure 14.5 *Showing the output from a noise source 25dB above thermal via an adjustable variable attenuator, against the setting of the attenuator.*

noise figures and other purposes, could be constructed for little more than the cost of the variable attenuator, or no more than a few pounds if an attenuator is already to hand.

References

1. Hickman, I. 'Making a right white noise', *Electronics World and Wireless World*, March 1992 pp. 256 and 257.
2. O'Dell, T. H. 'Self-calibrating noise source using silicon', *Electronics World and Wireless World*, Feb. 1993 pp. 170–172.
3. Hickman, I. 'Negative approach to positive thinking', *Electronics World and Wireless World*, March 1993 pp. 258–261.
4. Noise/Com, E.49 Midland Avenue, Paramus, New Jersey, 07652, USA. UK agent: Densitron Microwave Ltd, 4 Vanguard Way, Shoeburyness, Essex, SS3 9SH. Tel. 01702-463440, Fax 01702-293979.
5. Mini Circuits, a division of Scientific Components Corporation, PO Box 350166, Brooklyn, New York, 11235-0003, USA. In UK: Mini Circuits, Dale House, Wharf Road, Frimley Green, Camberley, Surrey GU16 6LF.

15 Ready to use RF amplifiers

Traditionally, an RF amplifier stage consisted of a suitable high frequency transistor, plus various discrete components for coupling, decoupling and biasing – resistors, capacitors and sometimes an RF choke. Additionally, for a tuned stage, a resonant circuit of some kind is used. Now, RF ICs incorporating one or more transistors and complete with biasing arrangements are available. But coupling capacitors and a feed resistor or RF choke are still needed.

In the design of an equipment destined for mass production in huge quantities, where every tenth of a penny on the component cost counts, whilst (thanks to surface-mount automated production machines) labour costs are minimal, it may pay to design a clever RF amplifier stage; especially if other special restraints, such as a current consumption measured in microamps rather than milliamps, apply. But for less demanding applications, and especially where time to market is more critical than component cost, an off-the-peg solution can be extremely attractive. In these circumstances, the ready-to-serve RF amplifiers which are described below can fill the bill perfectly. The amplifiers I refer to are the MAR series from Mini Circuits (a Division of Scientific Components Corporation) and the performance offered by the various members of the family is summarised in Table 15.1. It is an open secret that these are basically Avantec devices, but bought in huge quantities by Mini Circuits and sold at ridiculously low 'supermarket' prices, making them an extremely attractive buy.

Having purchased ten samples of each of three of the seven types available, it was decided to play with them to gain familiarity in use, before designing some of them into the first application I had in mind – a DDS-based signal generator which was under construction. One way to take a quick look to see what an amplifier can do is to connect its output back to its input, to implement an oscillator. As Figure 15.1 shows, the integrated two-stage amplifier is inverting, the component values having

Table 15.1 *Performance summary of MAR series amplifiers. (The colour dot referred to, in addition to denoting the type number, also indicates the input lead. This is opposite the output lead, the other two being ground.)*

Model No. Color Dot	FREQ. MHz	GAIN, dB Typical (at MHz) 100	500	1000	2000	MIN (Note 4)	MAXIMUM POWER, dBm Output (1dB) Compression	Input (no damage)	DYNAMIC RANGE Intercept pt. dBm NF dB Typ.	3rd Order Typ.	VSWR In	Out	MAXIMUM RATING (25°C) I(mA)	P(mW)	DC POWER at Pin 3 Current (mA) Typ.	Volt. Typ.
MAR-1 Brown	DC-1000	18.5	17.5	15.5	—	13.0	0	+10	5.0	15	1.5:1	1.5:1	40	100	17	5
MAR-2 Red	DC-2000	13	12.8	12.5	11	8.5	+3	+15	6.5	18	1.3:1	1.6:1	60	325	25	5
MAR-3 Orange	DC-2000	13	12.8	12.5	10.5	8.0	A+8	+15	6.0	23	1.6:1	1.6:1	70	400	35	5
MAR-4 Yellow	DC-1000	8.2	8.2	8.0	—	7.0	+11	+15	7.0	27	1.9:1	2.1	85	500	50	5
MAR-6 White	DC-2000	20	19	16	11	9	0	+15	2.8	15	2:1	1.8:1	50	200	16	3.5
MAR-7 Violet	DC-2000	13.5	13.1	12.5	10.5	8.5	+4	+15	5.0	20	2:1	1.5:1	60	275	22	4
MAR-8 Blue	DC-1000	33	28	23	—	19	+10	+15	3.5	27	□	□	65	500	36	7.5

been carefully designed to give a nominal match to 50Ω at both input and output (type MAR8 excepted). Thus it will oscillate at a frequency F_o when its output is connected back to its input via a length of 50Ω coax whose electrical length is $\lambda/2$ at F_o, as shown in Figure 15.2(a). With the length of coax shown and assuming it has a wave velocity of 65% that of free space, the expected frequency is

$$0.65 \times 300/(2 \times .965) = 101\text{MHz}$$

As the loss in the feedback network is negligible, the excess loop gain is virtually equal to the stage's forward gain, so the waveform would not be expected to be very good, as Figure 15.2(b) confirms. Given that the 1ns wide spikes on the edges of the waveform are way beyond the 250MHz bandwidth of the oscilloscope, their true amplitude must be more horrendous than it appears. The hard limiting in the amplifier is also responsible for excess phase delay (at this frequency the device exhibits 9° of excess phase anyway, even under small-signal conditions) or, put another way, the circuit is almost a relaxation oscillator, which always results in a lower frequency of oscillation than if the loop gain barely exceeds unity. In consequence, the actual frequency of oscillation is less than 100MHz, Figure 15.2(c), which shows high amplitudes of harmonics: the tenth harmonic is as large as the second, both barely more than 20dB down on the fundamental, whilst the third harmonic is only 8dB down – definitely not a clean oscillator.

Figure 15.3(a) shows the interesting effect of winding the supply voltage to the circuit of Figure 15.2(a) down from +12 to +8V. The narrow spikes are no longer so evident, but circuit behaviour is beginning to be chaotic. It exhibits a tendency to frequency halving (clearly shown in the frequency domain in Figure 15.3(b) – the circuit has reached the first bifurcation point.[1] A further slight change in supply volts prompts even more chaotic

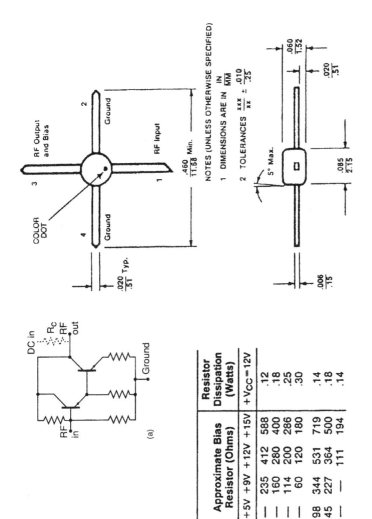

Figure 15.1 (a) The internal circuit of an MAR series amplifier. The resistor R_c is not part of the device, but provides an external DC feed path, whilst doubling as an RF choke. (b) An RF choke may be advisable in addition, in those cases where the value of R_c is fairly low (i.e. with the lower supply voltages).

NOTES (UNLESS OTHERWISE SPECIFIED)
1 DIMENSIONS ARE IN $\frac{IN}{MM}$
2 TOLERANCES $\frac{xxx}{xx} \pm \frac{.010}{.25}$

Amplifier	Bias Current I_B (mA)	Bias Voltage $+V_O$	Approximate Bias Resistor (Ohms)				Resistor Dissipation (Watts) $+V_{CC}=12V$
			+5V	+9V	+12V	+15V	
MAR-1	17	~5	—	235	412	588	.12
MAR-2	25	~5	—	160	280	400	.18
MAR-3	35	~5	—	114	200	286	.25
MAR-4	50	~6	—	60	120	180	.30
MAR-6	16	~3.5	98	344	531	719	.14
MAR-7	22	~4	45	227	364	500	.18
MAR-8	36	~8	—	—	111	194	.14

(a)

(b)

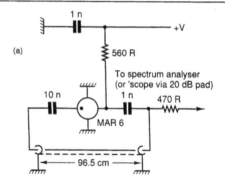

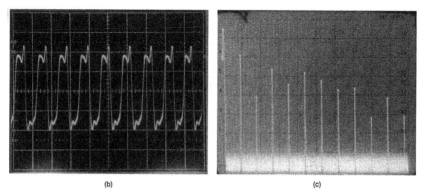

Figure 15.2 *(a) Connecting the amplifier's output back to its input via a halfwavelength of coax causes it to oscillate. In view of the wildly excessive loop gain, this circuit is sheer cruelty to helpless ICs. (b) The waveform produced by the circuit of (a). (10mV/div. vertical, 10ns/div. horizontal.) (c) The output of the oscillator, viewed in the frequency domain. (Ref. level +10dBm (but effectively higher due to the 470Ω resistor), vertical 10dB/div., horizontal 100MHz/div., 1MHz IF bandwidth, video filter OFF.)*

behaviour, with wide noise-like sidebands appearing around the fundamental and harmonics, Figure 15.3(c).

To produce a more sanitary oscillator, attenuation was added in the feedback loop, to reduce the excess loop gain, Figure 15.4(a). With R = 220Ω, the circuit did not oscillate, but did so with R = 150Ω. The amplitude control loop of Figure 15.4(b) was grafted on and the performance was then as illustrated in Figure 15.4(c). The output is very clean; second and third are the only significant harmonics, both well over 30dB down. The circuit of Figure 15.4(a) was then run without the amplitude control loop and with a 2–10pF trimmer in parallel with the 51Ω resistor in the feedback network. This provided a 7MHz tuning range, and it was noticeable that at the 10pF setting, the harmonics were substantially lower

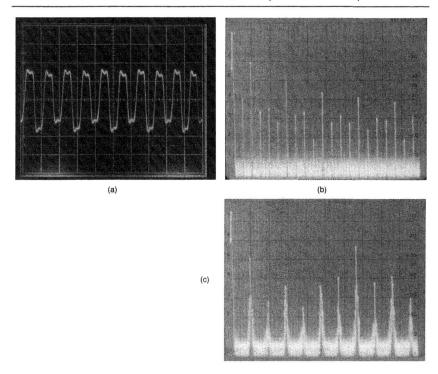

Figure 15.3 *(a) Showing the effect of winding the supply voltage to the circuit of Figure 15.2(a) down from +12 to +8V, frequency halving is evident. (b) As (a), in the frequency domain. (c) With further change of supply volts, behaviour is even more chaotic.*

than at 2pF, due to filtering action on the feedback signal; the levels of second to fourth harmonics seen were respectively −22dB, −28dB, −48dB, higher harmonics being negligible.

To check the performance obtainable at much higher frequencies with such a crude and simple oscillator, the circuit of Figure 15.4(a) was used, with the trimmer removed and the length of line drastically reduced. The circuit oscillated at 930MHz (Figure 15.4(d)) which shows how well the gain of the MAR6 is maintained with frequency, since R was still set at 150Ω. Note, however, that the amplitude was substantially reduced. The frequency of oscillation was well below that predicted by the line length, due to the excess phase shift through the amplifier at this frequency, amounting to some 70°, according to the data sheet. The spectral purity of the simple oscillators described above, with their frequency controlled by a length of transmission line, will not of course compare with that obtainable with an oscillator controlled by a high Q-tuned circuit. This is because in the latter case, the change of phase shift around the loop with change of

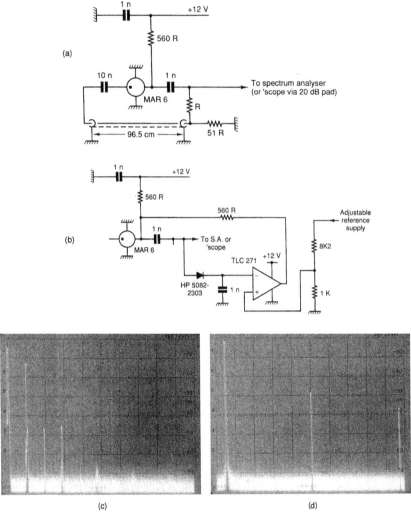

Figure 15.4 *(a) Circuit of Figure 15.2 modified to reduce the excessive loop gain. With R = 150Ω the circuit oscillates, but not with R = 220Ω. Second harmonic 30dB below fundamental, third harmonic 20dB down. The reduced output loading due to R enables the amplifier to supply almost its rated output power to an external 50Ω load circuit. (b) An add-on to (a), to define the amplitude of oscillation. (c) Perform-ance of (a) plus (b), showing the clean output obtained. Second and third are the only significant harmonics, both well over 30dB down. (Ref. level 0dBm, other settings unchanged.) (d) Performance of (a) alone with reduced line length, giving an operating frequency of 930MHz. (Ref. level 0dBm, centre frequency 930MHz, 100MHz/div. horizontal, other settings unchanged.)*

frequency is much more rapid than with a $\lambda/2$ transmission line. Figure 15.5(a) shows a possible configuration with the necessary 180° phase reversal provided by a tuned circuit, provided with matching for both ports of the amplifier. Even greater stability and spectral purity will result from crystal control and an 85MHz crystal was connected into the circuit of Figure 15.5(b). Being an overtone crystal operating at series resonance, it cannot conveniently be arranged to prove a phase reversal in the same way as a parallel resonant crystal, so a small two-hole balun core was used to provide the phase reversal. It was also arranged to step up the impedance presented by the crystal circuit to the amplifier's output, whilst a series-tuned circuit set to resonate at the desired frequency was interposed between the amplifier output and the reversing transformer, to suppress oscillations at any but the intended overtone. Excess loop gain was avoided by fitting a pad between the other side of the crystal and the amplifier's input. As Figure 15.5(c) shows, the circuit produced an output of −7dBm with low harmonic content, the waveform being shown in Figure 15.5 d). This is visibly a bit 'secondish', not so very different from an asymmetrical triangular wave. By taking the output from the point shown, it has not had the benefit of the filtering action of the frequency selective components. An output with lower harmonic content could be obtained from an additional buffer connected downstream of the series-tuned LC circuit, or even from the pad downstream of the crystal.

The experiments performed show that these amplifiers are delightfully tame and easy to apply, provided the two earth leads are connected directly to a ground plane. Microstrip construction is recommended with all transmission lines and ground plane running flush to the package, which means mounting it in a hole in the board. But for the experiments reported here, a fairly cavalier approach was adopted. The device was sat on top of the ground plane and the two ground leads were cranked down to connect to it, resulting in lead lengths of a millimetre or more, whilst other components were mounted in fresh air on 'skyhooks'. Even so, no problems of unwanted instability were encountered. The results show that when the amplifiers are run at well below the 1dB compression point, second harmonic distortion predominates (e.g. Figure 15.5(c) whilst of course with overdrive resulting in heavy compression, third harmonic is the largest, as the waveform approaches a squarewave, Figure 15.2(b) and (c). Where it is desired to obtain more output than available from a single device whilst retaining low harmonic levels, MAR series amplifiers can be paralleled as in Figure 15.6(a) – this is possible since they are (MAR8 excepted) unconditionally stable. The input and output impedances of the paralleled amplifiers fall within the range that is conveniently accommodated by standard 4:1, 9:1 and 16:1 broadband line transformer configurations. The bandwidth of such a paralleled stage will be limited by the bandwidth of the necessary matching transformers. In narrower

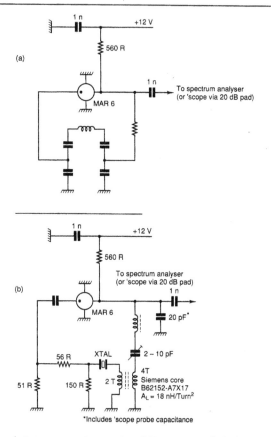

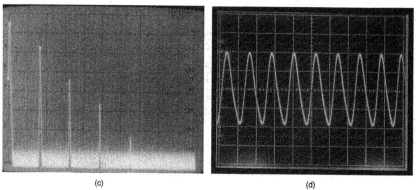

Figure 15.5 *(a) This circuit, with frequency determined by a tuned circuit, should offer improved spectral purity, i.e. reduced close-in noise sidebands. (b) Circuit modified for control by a series resonant overtone crystal. (c) Output of 85MHz crystal-controlled oscillator. (Ref. level +10dB, 50MHz/div. horizontal, 0Hz at LHS, other settings unchanged.) (d) Output waveform of the 85MHz crystal oscillator.*

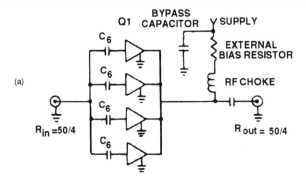

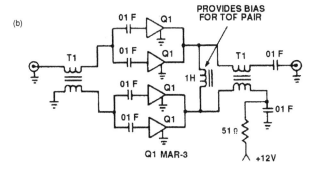

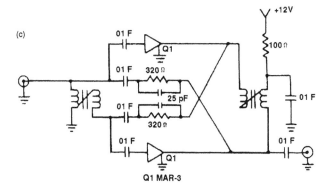

Figure 15.6 *(a) Being unconditionally stable, MAR series amplifiers may be simply paralleled for higher outuput power. (b) A parallel push-pull arrangement offers lower even-order harmonic output. (c) A push-pull stage may be unilateralised for greatly increased reverse isolation.*

bandwidth applications, other matching and combining techniques, such as quarterwave transformers and n-way Wilkinson splitters/combiners, can be considered. Note that the gain of such a compound amplifier is only the same as that of the component individual amplifiers, so to get an increased output, additional drive power must be applied. Paralleled amplifiers offer only increased output power, not lower levels of harmonics (unless derated). The four amplifiers in Figure 15.6(a) could advantageously be redeployed into the slightly more complex circuit of Figure 15.6(b) (which appears to have the 'μ' symbol missing from some component values!). In this circuit, owing to the push-pull arrangement, even-order harmonics will tend to cancel out.

A push-pull pair of MAR series amplifiers also lends itself to unilateralisation, Figure 15.6(c). (Unlike neutralisation, in which only the reactive components of a device's internal feedback are cancelled, unilateralisation is a technique in which both the real and imaginary terms of the feedback elements are cancelled. Consequently, unilateralisation tends to be effective over a wider frequency range than neutralisation.)

The reason that these amplifiers are so easily and effectively unilateralised is that the Q of their internal feedback paths is low compared to

(a)

Configuration	Freq. (MHz)	Gain (dB)	P_{-1dB} (dBm)	2nd Harmonic @ P_{-1dB} (dB below carrier)
Single-ended	100	12	+10	−15
Push-pull (Unilateralized)	100	15	+13.5	−26
Push-pull	100	12	+17	−34

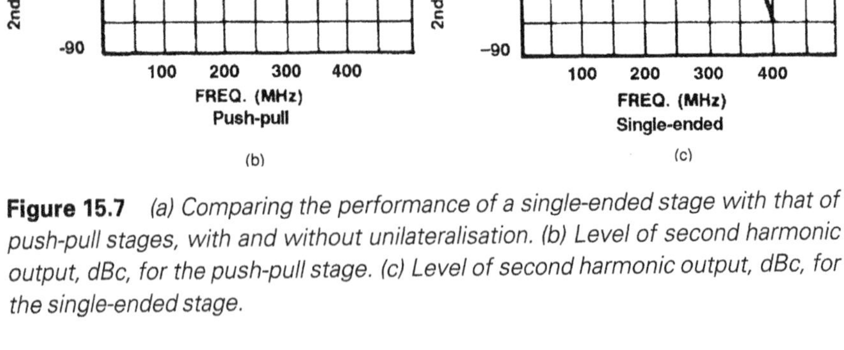

(b)

(c)

Figure 15.7 *(a) Comparing the performance of a single-ended stage with that of push-pull stages, with and without unilateralisation. (b) Level of second harmonic output, dBc, for the push-pull stage. (c) Level of second harmonic output, dBc, for the single-ended stage.*

conventional amplifiers. In an amplifier that has been unilateralised, the reverse isolation is greatly increased, so that variations in the load impedance will no longer affect the input impedance nor variations in the source impedance affect the output impedance. But unilateralisation tends to increase both the amplifier's input and output impedances, so careful attention must be paid to the effects of unilateralisation on input and output match.

Figure 15.7(a) compares the performance of a single-ended MAR series amplifier with that of a push-pull pair with and without unilateralisation. It is clear that the only major advantage of unilateralisation is the increased reverse isolation, the straightforward push-pull pair being better on other counts. In particular, unilateralisation has lowered the 1dB compression point by 3.5dB. This is partly due to the power lost in the resistive components of the cross-coupled feedback networks and partly to the effect on input and output impedances. Figure15.7(b) and (c) compares the level of second harmonic in dBc for the single-ended and push-pull amplifiers respectively.

Acknowledgements

Some of the diagrams and other material shown here are reproduced by courtesy of Mini Circuits. For further information contact Tel. 01252 835094.

References

1. Dettmer, R. 'Chaos and engineering', *IEE Review*, Sept. 1993 pp. 199–203.

16 Synchronous oscillators: Alternative to PLL?

> The synchronous oscillator is a fascinating circuit to play with, offering as it does an alternative approach from the PLL for signal extraction in a noisy environment. The possible variations on the theme are endless.

The phase locked loop (PLL) is a tried and tested circuit, well understood, capable of extracting signals buried in noise, and the usual crop of papers on the subject appeared at the latest yearly ISSCC International Solid State Circuits Convention. However, it is not the only way of pulling out a wanted signal from noise and (as one correspondent put it) the synchronous oscillator 'was vociferously advocated from the floor'. The synchronous oscillator was first brought to my attention by Ref. 6 and the ISSCC reporter's comments reminded me that I had intended to investigate the circuit further.

It is well known that an oscillator can be synchronised with an external signal of the same (or very nearly the same) frequency. This applies to oscillators of all sorts so that, for example, a Wien bridge-based audio oscillator can be locked (over a small range of frequencies) to a signal injected into the maintaining amplifier along with the circuit's own internal feedback signal. The injected signal does not need to be a sinewave, the oscillator will lock to a low-level squarewave whilst supplying its normal sinewave output, thus acting as a high Q filter. At RF, a weakly oscillating LC oscillator can be used in a similar way – a scheme which under the name 'reaction' (or, in the USA, the more colourful term 'tickling') is as old as the hills. Where the synchronous oscillator differs from this scheme is in the means by which the external signal, to which it is desired to lock the oscillator, is injected.

Figure 16.1 shows the basic circuit of an SO (synchronous oscillator) as it appears in Ref. 1 and several other of the referenced papers. Here, TR_2 is

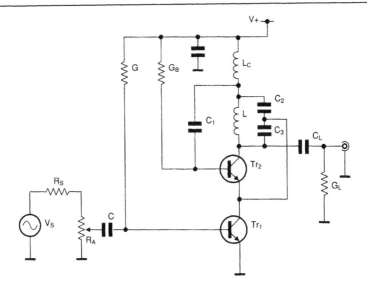

Figure 16.1 *Basic synchronous oscillator circuit as it appears in Ref. 1.*

arranged as a Colpitts oscillator, with its emitter current supplied from TR_1's collector. In the absence of any external synchronising input, the oscillator runs at the frequency determined by its tank circuit. An applied external signal modulates TR_1's collector current to a lesser or greater extent, even chopping it up into pulses in the case of a large applied signal. This modulation will synchronise the oscillator if the injected signal is close enough to the free-running frequency, though with a standing phase difference between the injected signal and the oscillator's output, of up to ±90° over the extremes of the frequency range for which lock is maintained. The stated theory of operation points out that the oscillator transistor TR_2 is designed to run biased well into class C, with a small conduction angle, with the result that it is 'blind' to any noise accompanying the externally applied signal for most of each and every cycle.

Anyone used to analog design will instantly see some shortcomings in Figure 16.1, in particular the very poorly defined DC conditions, due to the use of high resistance bias sources for two devices in series. The author of Ref. 1 states that the circuit is set up so that the transistors have equal collector-emitter voltages and thus both run in a linear regime. Whilst it is possible to select transistors or bias-resistors or both to achieve this, even with transistors of the same type tracking with varying temperature and ageing cannot be guaranteed, whilst some of the circuits in the referenced papers actually use different transistor types at TR_1 and TR_2. From Ref. 1 it appears that the circuit is run from +5V, which leaves only 2.5V V_{ce} for each device, assuming the biasing is perfectly balanced. Thus the maximum achievable tank circuit amplitude is about 4V pk-pk and the amplitude at the emitter of TR_1 will typically be a third of this. So the conduction angle will in fact be significant.

The circuit of Figure 16.2 was therefore chosen for initial experiments. Here, the DC conditions are well defined, with the oscillator transistor's emitter current supplied from one half of an LTP (long-tailed pair). With the tank components shown, the free-running frequency was 893kHz. Figure 16.3(a) shows the input waveform (lower trace) and the output (tank circuit) waveform with a signal input of 1V pk-pk at 875kHz (the lower end of the lock range) where the output is leading the input by 90°. Figures 16.3(b) and (c) show the same at 893kHz (input and output in phase) and at 919kHz (output lagging by 90°) respectively. Figure 16.3(d) shows the output of the SO with a swept input frequency. To maintain the light circuit loading provided by the 'scope probe, the spectrum analyser was not connected to the circuit directly, but driven from the oscilloscope's Channel 2 signal output, which provides a 50Ω output impedance. The input signal was swept from just above the lower limit of lock to just above the upper limit. At this point numerous FM sidebands appear, mainly on

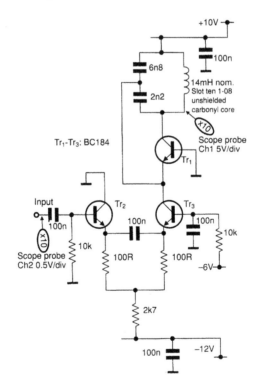

Figure 16.2   *Synchronous oscillator circuit used for initial experiments.*

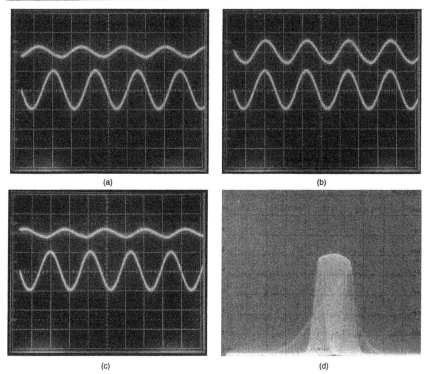

Figure 16.3 *(a) Input at 875kHz (lower trace, 0.5V/div.) and output (upper trace, 5V/div.) of a synchronous oscillator whose centre frequency is 893kHz, showing the output leading the input by 90° (0.5μs/div. horizontal). (b) As (a) but at the 893kHz centre frequency. (c) As (a) but input at 919kHz, output lagging by 90°. (d) Showing the output when the input signal was swept from just above the lower limit of lock to just above the upper limit. At the latter, a whole range of out-of-lock sidebands appear. (Vertical, 10dB/div.; horizontal, 20kHz/div., centre frequency 880kHz; IF bandwidth 3kHz, video filter off. Input to LTP -10dB ref 1V pk-pk.)*

the low frequency side, related to the rate at which the SO slips cycles in an unsuccessful attempt to retain lock. Note that the amplitude of the oscillator's output is a maximum at the centre of the lock range, i.e. at the frequency at which it free-runs in the absence of an input. Ref. 6 states that this is a characteristic of an injection-locked oscillator, whereas in a true SO, the output amplitude is constant over the whole lock range. Why the Figure 16.2 circuit exhibits this departure from the expected SO performance I am unable to say.

Another of the synchronous oscillator's party tricks is its ability to lock to an input frequency which is an integer multiple (harmonic) of its output frequency or to a submultiple. This is shown in Figures 16.4(a) and (b) respectively, again using the Figure 16.2 circuit. In each case, the input

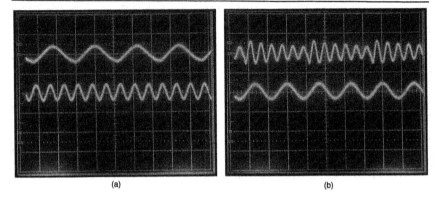

(a)   (b)

Figure 16.4 *(a) The SO of Figure 16.2 operating as a divider, synchronised by an input (lower trace) at three times its output frequency. Scope settings as Figure 16.3(a). (b) The SO of Figure 16.2 operating as a multiplier, synchronised by an input (lower trace) at one third of its output frequency. Scope settings as Figure 16.3(a) except 1µs/div. horizontal.*

signal amplitude was −10dB ref. 1V pk-pk. In (a), the 890kHz output (upper trace) is locked to an input at three times that frequency, i.e. the circuit is operating as a divider. In (b), the 890kHz output is locked to an input at one third of that frequency, i.e. it is acting as a tripler. The apparent amplitude modulation of the output at one half of the input frequency is one of those little mysteries that crop up all the time in electronics. I would have pursued it further, but was by this time beginning to feel that in some respects, the circuit of Figure 16.2 departed rather from the basic SO philosophy. Thus, for example, unlike the SO of Figure 16.1, the circuit will not readily act as a doubler, quadrupler, etc., since the balanced nature of the LTP stage feeding the Colpitts oscillator results in a minimal level of even harmonics in its output. It was time to look at the SO circuit as published in Ref. 1.

A version of the basic SO circuit of Figure 16.1 was therefore constructed, with TR_1 and TR_2 both 2N918s (selected for equal h_{FE}), G = 220kΩ, G_B = 100kΩ, L = 14µH, C_1 = 4n7, C_2 and C_3 both 100p and L_c = 22µH. This gave an operating frequency of around 6MHz, with a +5V supply rail. Although the two transistors operated with approximately equal values of V_{ce}, the oscillator transistor was bottoming heavily, viz. not operating in a linear regime. The two bias resistors were raised first to 1M and 470k, then 2M2 and 1M, but still the oscillator bottomed. Clearly the operating dynamic resistance R_d of the tank circuit was rather high, despite the heavy loading set by equal values for C_2 and C_3, but other worries about the circuit were beginning to niggle. For instance, although G_B in conjunction with the h_{FE} of the oscillator transistor sets the mean or DC value of the base potential, RF-wise, the base voltage can flap about. TR_1

has a high output resistance and so will not define TR_2's emitter potential, while the presence of the RF choke means that the top end of the tank is not clamped to the supply rail voltage (RF ground). Is the RF choke really necessary, and if it is, what defines the RF voltage at the ends of the tank circuit (and hence the voltage delivered to G_L, which in my case was a 'scope probe)? Doubts about the choke were confirmed by Figure 8 of Ref. 1, where a 10μH RF choke is used in a circuit operating at 560MHz. The reactance of an ideal 10μH inductor at 560MHz is over 35kΩ and it would resonate with a capacitance of just 0.008pF. Thus the impedance of the (unspecified) 10μH RF choke at the circuit's operating frequency is any-body's guess. Incidentally, this 'improved' SO circuit has a bias source for the lower transistor consisting of 111mV from a 5kΩ source, so that in the absence of an input of considerable amplitude, the lower transistor will be cut off, supplying zero emitter current to the Colpitts oscillator transistor.

It was time to return to the drawing board, and the Figure 16.1 circuit was abandoned in favour of that in Figure 16.5(a). As in Figure 16.2, the oscillator transistor's emitter current is closely defined, whilst the emitter of the lower transistor is at AC ground as in Figure 16.1. With no external input, the circuit oscillated at about 6.5MHz, with 10V pk-pk at the collector of the upper transistor, a little over 2V pk-pk at the emitter, Figure 16.6(a). The double exposure also shows the 'scope inputs

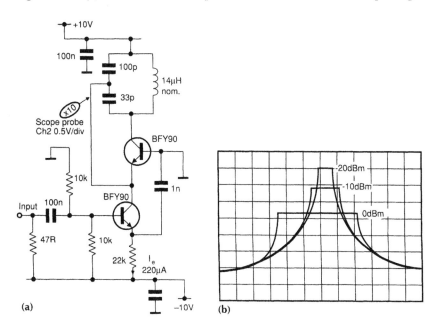

Figure 16.5 *(a) Modified SO circuit with robust definition of DC conditions. (b) Gain and bandwidth of a synchronous oscillator as reported in the published papers.*

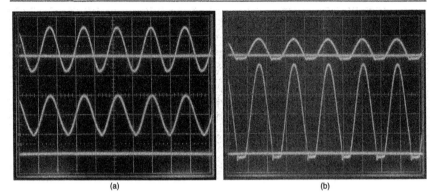

Figure 16.6 *(a) Waveforms associated with the SO of Figure 16.5(a). The emitter voltage reaches a peak of 0.8V negative with respect to the base, which is at 0V ground, as indicated by the superimposed grounded-input trace (upper trace, 1V/div.) The collector voltage (lower trace) swings 10Vp/p about the +10V rail, i.e. always at least 5V clear of the superimposed ground trace (lower trace, 5V/div.). Timebase 0.1μs/div. (b) As (a) but with too high an emitter current. These waveforms are typical of an oscillator circuit where a fixed base current rather than a fixed emitter current is used.*

grounded, showing the emitter reaching a peak negative-going excursion of −0.8V relative to ground (upper trace) and the collector voltage swing centred about the +10V rail. Thus the oscillator transistor operates with a narrow conduction angle and the collector voltage well clear of bottoming at all times. I hoped it would reproduce the performance shown in Figure 16.5(b), which indicates the gain of an SO versus input level as described in the references. Incidentally, Figure 16.6(b) shows the typical performance of an oscillator when the mean emitter current is not controlled to a level appropriate to the dynamic resistance of the tank circuit. Not only does the base emitter junction become forward biased, but the base collector junction does likewise at the negative extreme of its excursion. It thus appears as a forward-biased diode connected directly across the tank circuit, providing heavy damping which reduces the effective operating Q. The waveforms shown here resulted from a lower value of resistor (10kΩ) in the emitter of the lower BFY90 in Figure 16.5(a), but are typical of the result when an oscillator circuit provides the active device with a fixed base current. High gain samples of the oscillator transistor will try to pass more collector current than is appropriate, and the negative-going collector swing then mops up the excess base current, charging the associated base capacitors negatively in the process. The result is an overlong conduction angle, as shown.

Figure 16.7(a) shows the output of the SO of Figure 16.5(a) as a 0dBm input is slowly swept over a 500kHz range centred on 6.7MHz. At the start

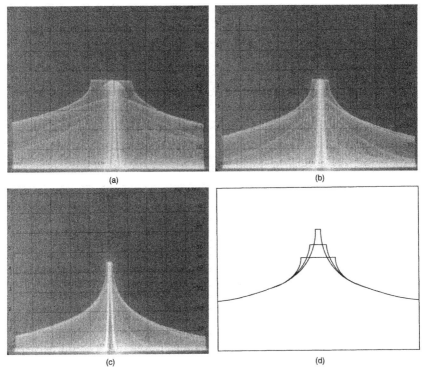

Figure 16.7 *(a) Output of the SO of Figure 16.5(a), level of input signal 0dBm. (Spectrum analyser settings: vertical, 10dB/div.; horizontal, 50kHz/div., centre frequency 6.7MHz; IF bandwidth 3kHz, video filter off. Spectrum analyser driven from oscilloscope's Channel 2 output.) (b) As (a) but −20dBm input. (c) As (a) but −30dBm input. (d) Traces (a)–(c) superimposed.*

of the sweep, the oscillator is not in lock and so runs at its natural centre frequency, producing a bright trace there, whilst the applied frequency can be seen to be well down the skirt of the tuned circuit. However, as the sweep proceeds, the amplitude of the applied signal does not trace out the shape of the tank circuit's response curve, but suddenly jumps up to the same level as the oscillation, as the latter synchronises with it. The lower level responses that are visible are due to the various out-of-lock sidebands seen in Figure 16.3(d): they would not have been visible if a spectrum analyser with a built-in tracking generator had been used. However, the top of the trace indicates the output at the wanted frequency, whether in lock or out. Figures 16.7(b) and (c) show the SO output with the input reduced from 0dBm to −20dBm and −30dBm respectively. The result with −10dBm was intermediate between (a) and (b).

Since the amplitude of the SO output in lock is independent of the level of the input, it follows that the smaller the input, the larger the 'gain' to the signal. Figure 16.7(d) shows tracings of the results in (a)–(c) superimposed such that the edge of screen levels are aligned, illustrating this point and giving the same sort of result as in the published papers. At 0dBm input, the lock range was 105kHz or 1.56% of the centre frequency, and 0.62%, 0.21% at −20dBm, −30dBm respectively. The acquisition of lock appears to be virtually instantaneous, unlike a second-order PLL using a simple ex-OR 'Type I' phase detector, where the time to acquire lock may run to many cycles of the input.

My tests to date have only covered the performance of the SO with a clean CW input – a very high SNR (signal-to-noise ratio) – whereas one of its main uses is said to be recovering a signal accompanied by noise. It seems that they are not ideal for recovering signals deeply buried in noise, Ref. 4 stating that the types experimented with and employed in various digital radio subsystems are not used below $E_b/N_o = -3$dB. I intend carrying out further tests in conjunction with a suitable noise generator, which will entail building a permanent version of the circuit reported in Ref. 7. It was in fact the hope that the SO would prove suitable for pulling out a signal with a large *negative* SNR that prompted the use of an LTP in Figure 16.2. With suitable emitter degeneration, the LTP would modulate the input signal plus noise onto the oscillator's emitter current in a linear manner. This is important, since with a large noise voltage applied to the LTP (and no degeneration) the signal plus noise input would in effect be hard limited. Now whilst hard limiting a signal well above noise improves the SNR by 3dB, hard limiting a signal buried well down in noise makes it 3dB worse. Perhaps this is the reason for the limited performance of the type of SO shown in Figure 16.1, where there are no specific measures to linearise the transistor forming part of the signal injection network. Ref. 6 quotes results with inputs up to 0dBm which, in a 50Ω system, corresponds to 636mV pk-pk. This is certainly more than the injection transistor (with no emitter degeneration or other linearising measures) can be expected to handle linearly. In fact, the transistor will start to bias itself back towards class C, so that both it and the oscillator transistor operate in a sampling mode. But with a negative SNR, this sort of limiting (by 'DC restoring' the signal negative going) probably risks suppressing the wanted signal energy even more than the symmetrical hard limiting produced by an LTP.

The SO is a fascinating circuit to play with, offering as it does an alternative approach to the PLL for signal extraction in a noisy environment. The possible variations on the theme are limitless. One difficulty it shares with the PLL when used for signal recovery is deriving a reliable in-lock indication. In a PLL used in a high SNR application, e.g. in a synthesiser, a lock detector often comes for free in the phase detector chip, along with an edge-triggered 'Type II' phase detector for rapid lock acquisition. However, in a PLL used for recovering a signal buried in noise, edge-triggered logic-machine type phase detectors have to be foregone, in favour of the 'Type I' ex-OR variety, with its smaller ±90° phase range. With this type of phase detector, the pull-in time can become quite long especially at poor signal-to-noise ratios. (The distinction between pull-in range and lock-in range is explained in Ref. 8, which further states that the distinction between the two can become rather blurred if appreciable noise is present in the circuit.) It will be interesting to see (posing a tricky measurement problem) if the SO still locks up instantly to a wanted signal buried in noise.

References

1. Uzunoglu, V. and White, M. H. 'The synchronous oscillator: a synchronisation and tracking network', *IEEE Journal of Solid State Circuits*, Vol. SC-20, No. 6, Dec. 1985. (With list of references to 20 other papers.)
2. Uzunoglu, V. and White, M. H. 'Some important properties of synchronous oscillators', *Proceedings of the IEEE*, Vol. 74, No. 3, March 1986, pp. 516–518.
3. Uzunoglu, V. 'Coherent phase-locked synchronous oscillator', *IEE Electronics Letters*, Vol. 22, No. 20, 25 Sept 1986.
4. Uzunoglu, V and White, M. H. 'Synchronous and the coherent phase-locked synchronous oscillators: new techniques in synchronization and tracking', *IEEE Transactions on Circuits and Systems*, Vol. 36, No. 7, July 1989.
5. Tam, M., White, M. H. and Zhigang, Ma. 'Theoretical analysis of a coherent phase synchronous oscillator', *IEEE Transactions on Circuits and Systems – 1: Fundamental Theory and Applications*, Vol. 39, No. 1, Jan. 1992.
6. Uzunoglu, V. 'The synchronous oscillator', *Electronic Engineering*, May 1993 pp. 41–47.
7. Hickman, I. 'Design brief "white noise – white knight"', *Electronics World and Wireless World*, Nov. 1993.
8. Gardner, Floyd M. *Phase Lock Techniques*, John Wiley and Sons Inc., 1966.

17 The ins and outs of oscillator action

The operation of many circuits is taken for granted. But this article explores the detailed functioning of the LC oscillator, to reveal unsuspected sophistication to its operation. In particular, a radical difference between the operation of conventional transistor oscillators and their vacuum tube predecessors is shown to be of fundamental importance in understanding their relative merits and limitations.

A good RF oscillator circuit should simultaneously fulfil a number of requirements, such as excellent medium- and long-term frequency stability, low harmonic content and low phase noise, to name but a few. These are not easy to achieve, even in a fixed frequency oscillator, and if an oscillator is tunable over a frequency range of more than a few per cent, it is even more difficult. Furthermore, in an oscillator covering an octave or more, it is desirable that the output amplitude should remain constant or at least very nearly so over the tuning range. In any case, if the amplitude varies widely with the tuning, the other parameters mentioned will be compromised.

I often wondered why it was that in many respects valve oscillators were so much better than the transistorised oscillators that replaced them. Clearly it had to do with the differences between a valve and a transistor. Comparing the grounded emitter circuit with the grounded cathode, the latter has a very high input impedance when, as is usual, the grid is negative with respect to the cathode, while the base input impedance of a transistor is by comparison, distinctly middling. Even when the grid of a valve is driven positive with respect to the cathode, its impedance is higher than that of a thermionic diode, as the grid – being wound of thin wire – does not make a very good anode. Comparing the collector and anode circuits, at DC and low frequencies the transistor presents a high output slope resistance, rather like a pentode, although considering internal feed-back via inter-electrode capacitances, the transistor is more like a triode. And there is one other very major difference between collector and anode circuits which is of paramount importance. When the voltage at the anode of a valve swings below the cathode voltage, the anode simply ceases to draw current. By contrast, when the voltage at the collector of an NPN transistor swings below that of the base, the collector/base junction becomes forward biased and when it swings below even the emitter voltage, the transistor works in the inverted mode where the collector acts as an emitter and vice versa. At one time, symmetrical transistors were manufactured, for use as crosspoint switches. Having identical emitter and collector structures, these devices worked equally well in either direction, although perhaps 'equally badly' would have been a better description. But modern transistors have very asymmetrical emitter and collector structures, and, being optimised for operation in the normal mode, they perform very badly in the inverted mode. In that mode, they present an impedance which might perhaps be described as a soggy mess, inflicting (in an oscillator) heavy resultant damping on the collector-tuned circuit. There is no reason why a diode in series with the collector could not be fabricated on the die. But it never is, at least not in small-signal or RF transistors, although you will find one in series with the drain of a 'Gemfet' or 'Comfet'. (These are proprietary names for the conductivity-modulated FET, in which injected minority carriers from the diode enhance the current-carrying capability and reduce the saturation voltage. A very handy device for the power engineer, but definitely not recommended for RF oscillators.)

Now a typical transistor oscillator circuit, such as the Hartley oscillator of Figure 17.1(a), is designed with a small-signal loop gain well in excess of unity, Figure 17.1(b)(ii). This guarantees that, when switched on, it will start to oscillate: nothing is more infuriating – and less useful – than an RF oscillator which works very well when running, but sometimes fails to get started at switch-on (Figure 17.1(b)(i)). But the excess loop gain at start-up has to be reduced somehow to a loop gain of just unity when running. In this type of single transistor circuit (as distinct from some other types of RF oscillator),[1] this is usually brought about by the collector voltage falling below that of the base. The collector/base junction thus becomes a forward-biased diode connected directly across the tuned circuit, imposing heavy damping upon it and thus reducing the loop gain by reducing the tuned circuit's effective dynamic resistance R_d. At the same time, the transistor, operating in the inverted mode, clamps the collector to ground, adding to the harmonic distortion in the output.

By contrast, a valve oscillator limits its amplitude in an entirely different way. Figure 17.2(a) shows a valve Hartley oscillator and Figure 17.2(b) shows the anode voltage and cathode current waveforms for different

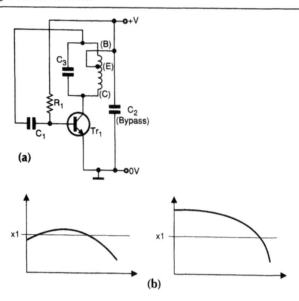

Figure 17.1   *(a) Basic bipolar transistor Hartley oscillator circuit. (b) Loop gain (Y axis) versus amplitude (X axis) of an oscillator which may fail to start (i), and of a reliable RF oscillator circuit (ii).*

degrees of loop gain, from so lightly coupled that the circuit barely oscillates, to heavily coupled with lots of excess loop gain. Given a short grid base, the valve works in class C with the anode current cut off for most of the cycle. At the negative peak of anode voltage, the positive peak of grid voltage rises above ground and grid current is drawn. The time constant C_1R_1 being long, the resultant negative average grid voltage has no time to discharge between cycles, the valve thus supplying its own negative grid bias. As the loop gain is increased, the peak cathode current increases and the peak to peak anode voltage swing rises until the valve bottoms on negative-going peaks. At this point, the cathode current cannot rise any further, however positive the grid becomes, but the current just either side of the negative peak can still increase somewhat. With heavy coupling, the anode voltage can swing below ground, but the points of the cycle where the valve feeds energy to the tuned circuit to maintain the increased swing are confined to the two regions either side of the negative peak, where the grid voltage is still near its positive peak but the anode is not bottomed. The anode current breaks up into two completely separated pulses, being zero in between. With further increase in amplitude, the anode swings further and further below ground and the two current pulses move further apart. They thus occur at a part of the cycle where the rate of change of anode voltage is greater; hence the time from grid voltage rising above cut-off to anode voltage falling below ground becomes shorter, strangling off the

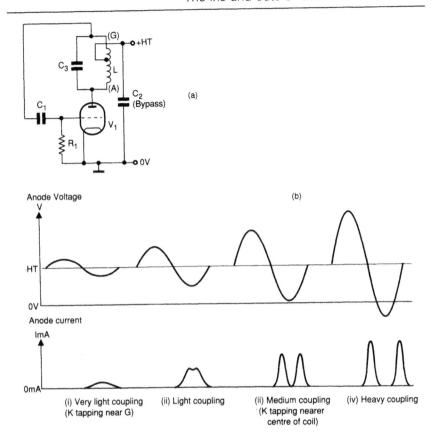

Figure 17.2 *(a) Valve Hartley oscillator circuit. (b) Anode voltage and cathode current waveforms for varying degrees of loop gain.*

current pulses to a narrower width. This reduces the component at the fundamental available to make up the tank circuit losses, leading to an equilibrium at a particular amplitude.

Many years ago I made up a test circuit to see if it were possible to simulate some of the features of a valve in a transistor oscillator circuit. Having only the most rudimentary equipment at the time, a low operating frequency, 20kHz, was chosen, enabling circuit operation to be easily viewed. Starting with the circuit of Figure 17.1(a), a resistor was added to the base circuit, to raise the device's input impedance to something nearer that of a valve's grid when forward biased. Then, a diode was connected in series with the transistor's collector, to prevent it conducting when its potential fell below that of the base. The completed circuit, Figure 17.3, drew 30mA from the supply and produced what appeared on an oscilloscope to be a perfect sinewave, swinging many volts below ground at the collector, despite the

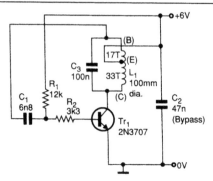

Figure 17.3 *Circuit of a low distortion 20kHz LC valve oscillator look-alike, using a transistor (see text).*

undoubtedly low Q of the coil (the R_d of the tank circuit was probably only of the order of 500Ω). Some small distortion was, however, observable on the smaller waveform at the base end of the tank circuit.

Being now better equipped, I decided to repeat the experiment at a higher frequency, but not so high that it would be impossible to observe the narrow current pulses expected. Also, to use a tunable oscillator to see how much the output amplitude varied across the tuning range. A tank circuit of 10µH (nominal) tuned by a 365pF (maximum) variable capacitor was therefore chosen, giving a lowest frequency of 2.5MHz. Note that over an octave tuning range, the R_d of the tank circuit will vary by about 2 : 1, and so therefore, to a first approximation, will the loop gain. If the collector current were constant, then a 2 : 1 variation in output amplitude could be expected. The intention was to use a JFET in place of a bipolar transistor, since the gate characteristic of this device resembles a valve, in that it normally draws no current, only conducting when driven above the source potential. A J310 was selected for the purpose, being an N-channel de-pletion VHF/UHF amplifier FET. Incidentally, this device has a typical equivalent short-circuit input noise voltage of just 10nV at 100Hz. Whilst this may be not too relevant in an RF amplifier, it is a definite plus point for an oscillator transistor, where the device's 1/f noise produces modulation sidebands about the output frequency, determining the level of the oscil-lator's very-close-in noise.

Alas, all attempts to use this device at the planned frequency were complicated by the J310's implacable resolve to oscillate at several fre-quencies simultaneously in the range 50–500 MHz, as well as performing (at first sight) as expected over a 2.5–5MHz tuning range. (My workbench has an extremely sensitive 418MHz receiver, part of a call system permit-ting the summoning of assistance from anywhere in the house or garden. The clacking of the needle in its RSSI meter alerted me to spurious radiation from the circuit under test, confirmed by an E field probe and the spectrum analyser.) The measured results at one frequency of an oscillator which was oscillating the whiles at other frequencies would clearly not be worth the paper they were written on, so the J310 was reluctantly abandoned.

A suitable lower frequency JFET not being to hand, a bipolar device was pressed into service. This was the BC182, with a minimum f_T of 150MHz, the particular sample used having an h_{FE} of 240. As with the J310, the circuit was constructed over a ground plane consisting of a sheet of copper-clad laminate, to which the frame of the tuning capacitor was firmly fixed. To permit grounding of the frame of the tuning capacitor, the Hartley circuit was modified to a tuned collector circuit with base feedback winding. A 4R7 resistor was placed in series with the BC182's emitter, to permit current monitoring. Initially, the inductor was grounded, the 4R7 emitter resistor being returned to a locally decoupled negative rail. However, it proved impossible to measure the small drop across this resistor due to imperfect negative rail decoupling and other causes, so the circuit was modified to use a positive supply as in Figure 17.4. From this it will be seen that in view of the higher operating frequency, the series resistor in the base circuit has been omitted, as it would have other effects and thus not well simulate the higher impedance of a valve grid circuit.

Figure 17.5(a) with its 100ns/div. time base shows the voltage at the anode of the diode at maximum tuning capacitance, a shade over 2.5MHz. With the 10V collector supply voltage, the 25V pk-pk voltage across the tank circuit results in the anode of the diode swinging well below the base voltage and indeed well below ground – 0V ground is at two divisions below the centreline, the upper trace at 5V/div., and both traces DC coupled. The other trace, at 50mV/div., is the voltage across the 4R7 emitter current sensing resistor, and it proved quite difficult to measure.

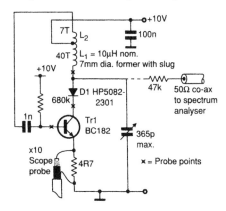

Figure 17.4   *Circuit of the 2.5–5MHz 'valve style' oscillator. The 47k connection to the spectrum analyser was removed when not in use.*

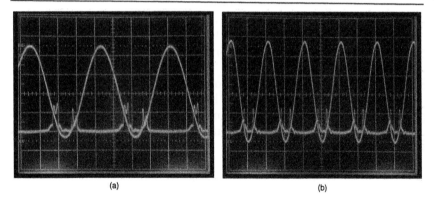

(a)   (b)

Figure 17.5   *(a) Tank circuit waveform with tuning capacitor set to max, 2.5MHz (upper trace, 5V/div.) and emitter current waveform (lower trace, 50mV/div.). Ground level two divisions below centreline, 100ns/div. horizontal. (b) As (a) but tuning capacitor set for 5MHz output.*

The magnetic field from the coil coupled with the probe's ground lead, however it was dressed. In the end, the probe ground lead was removed entirely and the probe's tip and earth ring strapped across the resistor body as indicated in Figure 17.4. As in a valve oscillator, the collector current has split, in this case due to the presence of the diode, into two separate pulses, each flowing only whilst the base is forward biased and collector voltage above the transistor's bottoming voltage. The ringing on these two pulses is possibly due to the inductance of the 4R7 resistor, and doubtless other circuit parasitics also, pointing up the wisdom of not attempting the experiment at too high a frequency.

Figure 17.5(b) shows the same picture, but with the circuit tuned to oscillate at 5MHz. Bearing in mind that the reactance of the inductor at 5MHz will have doubled relative to Figure 17.5(a)'s 2.5MHz, then assuming the were Q unchanged (only approximately true), the tank circuit's dynamic resistance would have doubled. Yet the amplitude of oscillation has increased by only a few per cent. The reason is that the collector current pulses are now very much narrower – not only in absolute terms, but, more importantly, as a fraction of a cycle. Thus the total conduction angle is reduced, and with it both the mean collector current and the component at the 5MHz fundamental. Whilst the peak amplitude of the pulses is little changed, they are now only a few nanoseconds wide. With the 15µA base current supplied and the device's h_{FE} of 240, the collector current drawn when the base feedback was removed, stopping the oscillation, was 3.6mA. At 2.5MHz it was 1.6mA and this fell further to 1.3mA at 5MHz. The mean base current was of course unchanged, the excess being spilled through the base circuit during the period when the collector current was zero due to the diode being reverse biased.

Figure 17.6(a) shows the output spectrum at 2.5MHz (span 0–20MHz), that at 5MHz being the same, except that the second harmonic rose to −32dBc: in both cases, harmonics higher than the fifth were negligible. The spectrum analyser's reference level (top of screen) is −10dBm, but due to the 1000 : 1 attenuation introduced by the 47K resistor, it corresponds to +50dBm − at least in terms of volts, though not in terms of power, of course, as the tank circuit impedance is much higher than 50Ω. Figure 17.6(b) shows the base voltage waveform at 2.5MHz (lower trace, 1V/div.) and a waveform (upper trace, 5V/div.) which could not be seen in the corresponding valve oscillator. This is the waveform at the cathode of the diode − although the anode of a valve oscillator acts as a diode, cutting off when its voltage falls below ground, there is no corresponding cathode available for monitoring. The collector can be seen to be firmly clamped to ground at the negative peak (when the diode is reverse biased), subsequently rising to the positive peak of the tank circuit voltage. Thereafter, it remains there until the transistor turns on again, at the first of the two current pulses surrounding the following negative peak.

Figure 17.7(a) shows the tank-circuit/collector voltage when the diode is short-circuited, to give conventional transistor LC oscillator operation. Here, the negative peak is brutally clamped to ground: compare this with Figure 17.5(b), where the tank-circuit voltage is free to swing 5V below ground. The extra damping has reduced the swing from Figure 17.5(b)'s 28V pk-pk to 25V pk-pk. The neat snipping off of the negative tip of the waveform does not affect the low order distortion greatly, but as Figure 17.7(b) (span 0–100MHz) shows, the significant harmonics now extend up

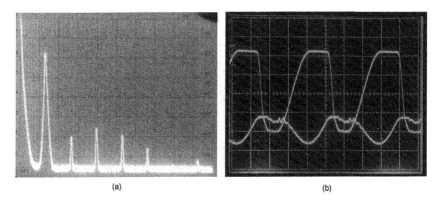

(a)   (b)

Figure 17.6   *(a) Spectrum of the output of the circuit of Figure 17.4 at 2.5MHz. Vertical 10dB/div., ref. level −10dBm, span 0–20MHz, IF bandwidth 100kHz, video filter on. (b) Waveform at the collector (cathode of the diode) at 2.5MHz (upper trace, 10V/div.) and the base (lower trace, 1V/div.), 0V level two divisions below centreline, 100ns/div.*

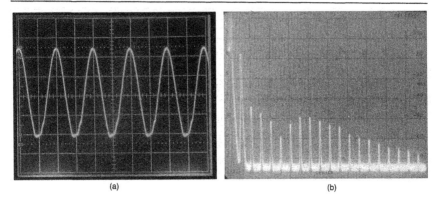

(a)   (b)

Figure 17.7   *(a) Waveform at the collector (tank circuit) with the diode short-circuited, at 2.5MHz (10V/div.), 0V level two divisions below centreline, 100ns/div. Tank circuit voltage cannot swing below ground. (b) Spectrum of (a). Vertical 10dB/div., ref. level −10dBm, span 0–100MHz, IF bandwidth 1MHz, video filter on.*

to a much higher order. Incidentally, the emitter current also breaks up into two pulses in this circuit, but for an entirely different reason from the case where the diode is present.

Nothing could show the difference between a conventional transistor LC oscillator and the 'pseudo valve' circuit better than Figure 17.8. Figure 17.8(a) shows the base voltage waveform of the pseudo valve circuit at 2.5MHz (at 0.5V/div.) and the emitter current pulses monitored across the 4R7 resistor (at 50mV/div.). Note that the base voltage stays positive

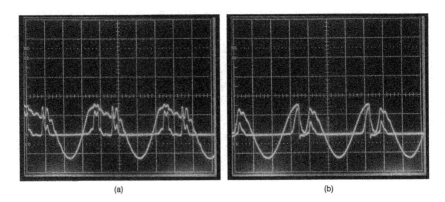

(a)   (b)

Figure 17.8   *(a) Pseudo valve circuit. Base circuit waveform with tuning capacitor set to max, 2.5MHz (larger trace, 0.5V/div.) and emitter current waveform (smaller trace, 50mV/div.). Ground level two divisions below centreline, 100ns/div. horizontal. (b) Conventional circuit. 2.5MHz. Traces and scope settings as (a).*

during the period between current pulses, when the tank circuit voltage is negative. This is in complete contrast to the conventional circuit without the diode. Here, Figure 17.8(b), when the collector tries to swing below ground the base-collector diode turns on, dragging the base voltage down with it. This reverse biases the base-emitter junction, interrupting the emitter current and splitting it into two separate pulses. In this circuit, the excess base bias current is disposed of into the collector circuit whilst the emitter current is off. In the pseudo valve circuit, it is disposed of into the emitter circuit, whilst the collector current is cut off.

The differences between a conventional transistor LC oscillator and the 'pseudo valve' circuit shown here can be expected to apply to the two circuits when operating at much higher frequencies. Some of the effects, such as the ringing on the emitter current pulses seen in Figure 17.5(a), are due to a poor RF layout (required to permit instrumentation), and the intrusion into the circuit of probes. In a practical application, these would not be present, and, given its advantages, the 'pseudo valve' oscillator could be seriously considered for applications at substantially higher frequencies.

References

1. Hickman, I. 'Design Brief "Oscillator tails off lamely?"', *Electronics World and Wireless World*, Feb. 1992.

18 The transformer-ratio-arm-bridge

Every so often, it becomes worthwhile to reinvent the wheel. This article speculates on a new type of instrumentation, based on the transformer-arm-ratio-bridge, a variant on the original Wheatstone bridge.

The use of the bridge principle in electrical measurements has been extended far beyond Wheatstone's original application for the measurement of resistance at DC. It has been adapted for measuring inductance (e.g. Hay, Anderson, Butterworth, Heaviside-Campbell and Maxwell bridges), capacitance (e.g. Schering bridge, often with Wagner earth, etc.) and even frequency (e.g. Wien bridge). Latterly, bridge methods have largely fallen into disuse, especially for RF measurements. The reason is that each measurement is taken at a spot frequency and involves adjusting two standards for balance, so that investigating the variation of impedance or admittance of an unknown with frequency is tedious, involving the plotting of a large number of spot measurements. It is so much easier to connect the unknown to a network analyser and take an s_{11} measurement covering the frequency range of interest. The answer can be viewed instantly as an $M \angle \phi$ plot versus frequency, or I and Q components, or as a Smith chart display, with the value corresponding to a marker at any desired spot frequency indicated on the screen as a numerical readout. With the convenience of such instant answers, it is no wonder that the RF bridge has been largely relegated to history.

Often the unknown is not a simple two-terminal impedance, but a two-port pi network, as in Figure 18.1(a). Given three independent equations relating the three unknowns to three measured values, the various impedances or admittances can be calculated. So any three of the four s parameter measurements in Figure 18.1(b) should in theory suffice, permitting the evaluation of the three impedances. In practice, if Z_1 and Z_3

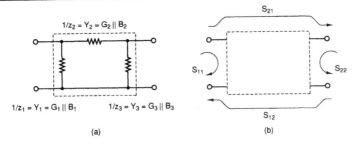

Figure 18.1 *(a) General two-port pi network expressed in parallel components. (b) s parameter measurement of two-port pi network.*

are low and Z_2 high, the computation will involve the difference of two large quantities, magnifying any measurement errors and may lead to a large margin of error for the calculated value of Z_2. This is precisely where a particular type of RF bridge, the transformer-ratio-arm-bridge, scores over other methods. The TRAB was developed during WWII by Mr. Gilbert Mayo of the BBC Research Department, and subsequently further developed and marketed by at least two companies in the UK. They were at one time available from Hatfield Instruments (Type LE300A, covering frequencies up to about 15MHz) and from Wayne Kerr (Types B800 and B801, covering – from memory – up to 100MHz and 250MHz respectively).

The transformer-ratio-arm-bridge

Figure 18.2(a) shows an RF TRAB in its simplest form. An RF test signal from a bridge source or signal generator is applied, via a step-down transformer T_1, to a calibrated variable standard capacitor C_s and a conductance standard G_s which is effectively variable from $1/R_1$ down to zero by means of the non-inductive calibrated variable resistor R_V. The other ends of these two components are connected to one end of a centre-tapped symmetrical winding on T_2. The unknown is connected between the output of T1 and the other end of the centre-tapped winding on T_2. An output winding on T_2 is connected to a bridge detector, which can be any radio receiver covering the band of interest. In use, C_s and G_s are adjusted until there is no detectable output at the receiver. The process of obtaining a balance can create rather a disturbance in a lab, as the receiver's AGC must be switched off – since with a tight AGC loop there is no indication which way to adjust C_s and G_s to approach balance! Balance occurs when the parallel components C_x and G_x of the unknown are equal to the capacitance setting of C_s (marked on its dial) and the effective value of G_s (which is marked on the dial of R_V). Balance occurs as a result of the

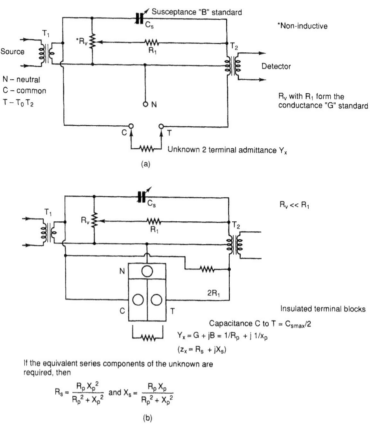

Figure 18.2 *(a) In its simplest form, a TRAB measures only capacitance and conductance (resistance). (b) With modifications, it can also measure negative capacitances (inductance) and negative conductances. Additional circuitry (not shown) is required to permit balancing of the bridge with C_s and R_v set to zero, before connecting the unknown.*

current flowing via the standards through one half of the balanced symmetrical winding of T_2 equalling the current flowing via the unknown through the other half of the winding. As the number of turns on each half winding are the same, there is no net magnetising force, and consequently no resultant flux on the core. There is thus no voltage induced in the output winding, but equally if not more important, there is no voltage appearing across either half of the centre-tapped winding either. Thus T_2 acts as a virtual earth, the significance of which will appear in a moment. Note that a TRAB inherently measures the unknown Z_2 in terms of an admittance Y_2, i.e. equivalent components G_x in parallel with B_x. Given the known test frequency, these can be converted to the equivalent series components R_x

and X_x using the formulae in Figure 18.2(b), or using a Smith Chart.[1]

As shown in Figure 18.2(a), the range of the unknown that the bridge can handle is limited to C_x not exceeding the maximum value of C_s, and G_x not higher than $1/R_1$, with no capability at all for measuring either inductance, or negative conductance. This capability can be provided by the modified arrangement of Figure 18.2(b). Here, the terminals to which the unknown are connected are mounted upon two substantial blocks of metal, firmly bolted together with an insulating film between, to form a capacitance equal to $C_{s\ max}/2$. The dial of C_s is now calibrated in positive and negative values of capacitance, with zero occurring where C_s is set at $C_{s\ max}/2$; this arrangement was employed in the Wayne Kerr TRABs. Similarly, a resistor of value $2R_1$ can be added in parallel with the unknown terminals, allowing for negative values of conductance and permitting measurements on active devices. The shunt components of resistance (or conductance) and capacitance of the unknown could be read directly from the dials, at any test frequency. Inductance, on the other hand, was read as a negative capacitance. Knowing the test frequency enables the susceptance of the negative capacitance, and hence of the unknown inductive component, to be calculated, from which the value of inductance itself was readily derived. Although shown with the source connected to T_1 and the detector to T_2, being an entirely passive linear network, a TRAB can be used with the source and detector interchanged, where convenient.

The Wayne Kerr bridges were much as described above. This meant that the range of conductance, capacitance and inductance that could be measured was limited, but the bridge circuitry and transformers could be kept simple and very compact. They were thus usable to VHF or low UHF. The Hatfield Instruments LE300A bridge, on the other hand, was designed to measure the widest possible range of unknowns, limiting its top frequency to 15MHz due to the more complicated transformers and the switching arrangements. (Another version, the LE300A/1, was available which enabled DC polarising voltages and currents to be introduced to the unknown without affecting the RF measurements.) The source was connected via one of the pins of the left-hand twin axial socket, the right-hand socket providing the output to the detector: when an internal source and detector were fitted, twin axial shorting plugs with the two pins strapped were used. As with the Wayne Kerr bridges, set zero controls were fitted, to enable balance to be achieved at any given frequency with the C_s and G_s dials set to zero, before connecting the unknown. The unknown could be connected to any one of three terminals, with multiplying factors of 10, 1 and 0.1. A special low-loss wafer switch connected the standard resistance R_4 to one of six tappings on T_2 – which actually consisted of two separate interconnected transformers – providing (in conjunction with the terminal multiplier) for the measurement of conductance with ranges from 0 to

±100mS full scale down to 0 to 0.01mS full scale. C_s was similarly switched, providing for the measurement of both capacitive and inductive susceptances in ranges 0 to ±2.5pF full scale up to 0 to ±25 000pF full scale. Note, however, that the bridge was designed to work with admittances of 100mS maximum, so that, for example, the maximum measurement frequency for a 25nF capacitor (where the susceptance was j100mS) was 670kHz. In addition to being calibrated in picofarads, the C_s dial was also directly calibrated in both microhenries and reactance, as applicable at the spot frequency of $\Omega = 10^{-7}$ rad/s.

Extended possibilities

Figure 18.3 shows how the TRAB is inherently adapted for three terminal measurements, permitting the accurate measurement of Y_2 whatever the values of Y_1 and Y_3, by virtue of the bridge's neutral connection. It can be seen that admittance Y_1 shunts the source whilst Y_3 shunts the detector (T_2 is a virtual earth). Thus whilst Y_1 in particular may reduce bridge sensitivity slightly, neither Y_1 nor Y_3 affects the accuracy of the measurement of Y_2. Transformer T_1 in Figure 18.3 provides a 3 : 1 step-down ratio, so that given a 50Ω source, the bridge is driven from an effective source resistance of less than 6Ω. Y_3 shunts the detector, but the voltage across it is negligible, and ideally zero if the leakage inductance between the two halves of the symmetrical winding is vanishingly small.

A range of special adapters were available enabling the LE300A to measure a variety of Y parameters of both NPN and PNP transistors, but in view of the limited frequency range of the bridge, these are now of only antiquarian interest. However, another type of adaptor was available which provided a most remarkable extension to the measurement range of

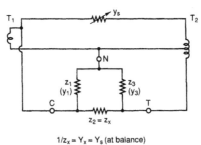

Figure 18.3   *Circuit showing how a TRAB is inherently perfectly adapted for three terminal measurements. Admittance Y_1 shunts the source whilst Y_3 shunts the detector (T_2 is a virtual earth). Thus whilst they may reduce bridge sensitivity slightly, neither affects the accuracy of the measurement of Y_2.*

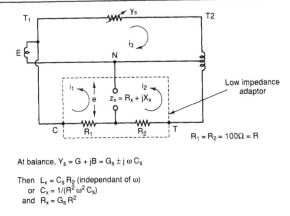

At balance, $Y_s = G + jB = G_s \pm j\,\omega\,C_s$

Then $L_x = C_s R_2$ (independant of ω)
    or $C_x = 1/(R^2 \omega^2 C_s)$
    and $R_x = G_s R^2$

Figure 18.4 *Showing how a low impedance adaptor not only extends the measurement range of a TRAB, but also turns conductance into resistance, and capacitive susceptance into inductive reactance and vice versa.*

the bridge – these were the low impedance adaptors type LE305 and LE306. The lowest impedance which could be measured on the LE300A TRAB directly was 10Ω and the purpose of the adaptors was to extend this down to 0.001Ω. Thus in addition to resistances down to that value, with the LE305, inductances down to 1nH could be measured. The maximum capacitance that could be measured with the adaptors depended upon frequency, rising from 4nF at 2.5MHz to 10 000μF at 15kHz.

Automated TRAB?

The facility for measuring directly the parallel components of susceptance of the series element of a pi network is so attractive that one is led to wonder if with today's technology one could produce a self-balancing TRAB. The drive signal could then be swept, to give an impedance versus frequency display, not unlike a network analyser. For instance, the variable conductance standard formed by R_1 and R_v in Figure 18.2 could be replaced by a four quadrant multiplier, with the RF applied to the X port and the control signal to the Y port. The output would thus be adjustable in both amplitude and sign, avoiding the need for selecting one side of T_2 or the other. The bridge output from T_2 would be synchronously detected and the resultant fed back to the Y port of the multiplier to automatically achieve a balance with the conductance of the unknown. The degree of balance achieved would depend upon the loop gain. The variable suscep-tance could be obtained from another multiplier instead of C_s, by feeding its X port from a quadrature version of the drive signal. This requirement is easily met since sine and cosine outputs are available as standard from

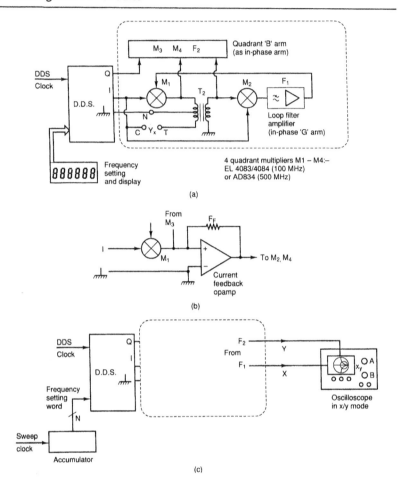

Figure 18.5 *(a) TRAB with autobalance facilities. (b) As the multipliers can imple-
ment RF outputs of both polarity, i.e. can act as both negative and positive
conductance and susceptance standards, T_2 is redundant, and can be replaced by
an op-amp virtual earth. (c) As (a), with sweep facility and Smith Chart display.*

many direct digital synthesizer chips. The arrangement might look something like Figure 18.5(a), where the synchronous detectors could be multipliers identical to those used as the G and B standards, fed through lengths
of line to compensate the path length through the bridge. As the multipliers
can implement RF outputs of both polarity, i.e. can act as both negative
and positive conductance and susceptance standards, T_2 is in fact redundant. It can be replaced by a current feedback op-amp virtual earth as in
Figure 18.5(b): conveniently the multipliers provide current outputs. Fig-

ure 18.5(c) shows how a swept measurement, with Smith Chart display of the unknown as a function of frequency, might be organised.

The only minor disadvantage of this 'active TRAB' arrangement is that a very low output impedance buffer is needed to drive the unknown. For whilst with this arrangement, any reduction of the I drive voltage due to loading by Z_1 (or indeed Z_2) will affect the conductance standard equally and thus cause no error, it will not correspondingly affect the susceptance standard arm.

References

1. *Electronics World and Wireless World*, Mar. 94, pp. 256–26 (Reproduced below in Part 4, Basic Principles, see p. 245).

19 Smaller steps to better performance

Active multipliers find many uses in instrumentation. With its wide bandwidth, the AD834 can find use in the modulation and levelling departments of a signal generator. The following article provides some suggestions on just how.

Amongst the writer's tally of professional and home-built test equipment there is – in the latter category – a DDS-based signal generator covering about 1 Hz–320 MHz in (approximately) 1 Hz steps. It provides only a CW output and is used in conjunction with an external 0–120 dB step attenuator providing 10 dB steps. Whilst the attenuator could have been built into the case of the signal generator, keeping it separate provides greater flexibility in use, enabling it to be used with other signal sources. The obvious lack in this arrangement is any facility for amplitude modulation or output level adjustments in steps finer than 10 dB. A sample of the Analog Devices AD834 500 MHz Four Quadrant Multiplier coming to hand, this seemed the ideal opportunity to experiment with the device while at the same time filling a long-felt want.

In commencing the design, certain constraints were allowed for: for example, although the output of thhe DDS signal generator was about 0dBm, there was a variation of nearly 2dB over the full range up to 320 MHz. Furthermore, it was clear that several RF amplifier stages would be needed in the RF path, with each contributing some further gain variation. It was therefore decided to enclose the whole RF path through the modulator/attenuator within a levelling loop. This has a further advantage. Since the leveller output is maintained constant regardless of input *or load* variations (within reason), it represents – effectively – a zero output impedance point. From thence, the load can be supplied via a 50Ω resistor, giving an ideal generator output impedance quite independent of the actual output impedance of the last RF amplifier in the chain.

The AD834 accepts a maximum differential input on both its X and Y balanced inputs of ±1V pk-pk, producing at its differential W output port a current of ±4mA full scale, according to the relation

$$W = \frac{XY}{(1V)^2} \cdot 4mA$$

The output of the DDS signal generator was about 0dBm or only some 630mV pk-pk, so some amplification was indicated in order to take full advantage of the multiplier's ±1V dynamic range. With frequencies up to 320MHz at the multiplier's input, balanced circuitry is not convenient, but fortunately the device's common mode rejection at both X and Y input ports is such that in each case one lead can be grounded and the other driven unbalanced, as indicated in Figure 19.1, which shows the whole RF path through the modulator/attenuator. As recommended in the data sheet when using unbalanced inputs, X_1 and Y_2 were grounded, the inputs being applied to pins 1 (Y_1) and 8 (X_2). Similarly, for convenience, an unbalanced output was taken from W_2, W_1 being returned to the supply, even though this halved the available pk-pk output current to ±2mA. This alternating current is superimposed upon a standing 8.5mA (nominal) DC component and is sunk by an open collector output. The open collector W outputs must be operated at a voltage slightly above that on pin 6 (V+). This was arranged following the manufacturer's recommended method of inserting a resistor in series with the supply to pin 6. The AC component, flowing in 47Ω load resistor R_6 in parallel with the (nominal) 50Ω input impedance of IC_3, forms the output voltage from the modulator. This was

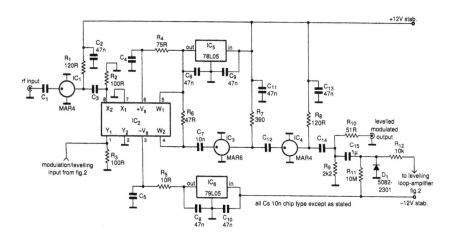

Figure 19.1 *The RF path through the modulator/attenuator/leveller.*

applied to the following amplifier stages. Note that the 390Ω supply resistor of IC_3 is somewhat lower than the recommended value, so the device dissipation will be increased, but this is acceptable for lab. use as distinct from a full-temperature-range application.

After further amplification, the signal voltage at the output of IC_4 corresponds to +6dBm when the voltage on the Y1 input of the modulator is rather less than +1V (or −1V). Thus the level delivered to a matched load at the output is just 0dBm. The RF output of IC_4 is DC restored positive going by D_1, whose linearity versus signal level is improved somewhat by a soupçon of forward bias via R_{11}. The mean level of the voltage at D_1's cathode is (almost) equal to the peak RF voltage, and it is applied via R_{12} to the modulator/attenuator/levelling loop, which is shown in Figure 19.2.

Whereas the circuitry of Figure 19.1 was constructed on a scrap of single-sided copper-clad laminate, used as a ground plane, that of Figure 19.2 was built on a piece of 0.1inch matrix copper stripboard. The level detector output is applied to the loop filter amplifier, IC_{7B} and associated components. IC_{7B} controls the gain of the transconductance amplifier stage IC_8, which receives its input from IC_{7D}. IC_8 is an LM13600, of which one half is unused. (An LM13700 will do just as well, there being only a minor difference between these two devices. In the former the

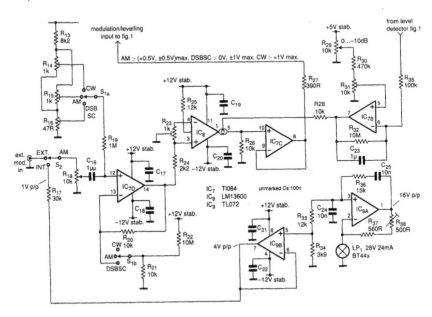

Figure 19.2 *The rest of the modulator/attenuator/leveller circuitry, including a 1kHz oscillator for the internal modulation facility (IC_9), the loop amplifier/filter IC_{7B}, the variable gain amplifier (OTA IC_8) and the carrier/modulation scaling stage IC_{7D}.*

emitter current of the input transistor of the Darlington output buffer is controlled pro rata with the g_m of the transconductance section, providing improved dynamic range, whereas in the latter it is fixed. Since in this application the output buffer is not used, either device will do.) R_{25} provides the current to operate the LM13600's input linearising diodes.

IC_{7D} produces a DC level which determines the level of the carrier at the output, combining it with an AC signal where modulation is required. For maximum CW output, IC_{7C} (which buffers the OTA's output) applies the required voltage (up to a maximum of +1V) to the modulator's Y_1 input, via R_{27}. This occurs with the 0 to 10dB attenuator control R_{29} set to maximum, R_{31} having been suitably adjusted. R_{29} provides an attenuation range of over 10dB, and although its operation is approximately linear rather than logarithmic, it can readily be calibrated with a dB scale. Operation on AM is similar, except that the DC level at the modulator's Y_1 input is halved, to allow for up to 100% modulation. An internal 1kHz modulation oscillator, IC_9, is included and the modulation depth can be set by R_{18}. R_{18} can be calibrated directly in percent amplitude modulation depth, the level supplied by the internal modulation oscillator being such that fully clockwise corresponds to 100% AM. If the internal modulation oscillator is run with a low output swing at IC_{9A} such as 4Vpk-pk, setting up is critical and amplitude control poor, due to inadequate drive to the lamp used to stabilise the loop gain. With the arrangement shown, giving 16Vpk-pk, control is tight with little amplitude bounce at switch-on. A 4Vp/p output is picked off by IC_{9B}, which it should be noted is driven from the frequency selective network, rather than the output of the maintaining amplifier. Thus IC_{9B}'s output has the advantage of the selectivity of the Wien network – this only amounts to 2.5dB at third harmonic relative to the fundamental, but every little is worth having. The measured total harmonic distortion at IC_{9B}'s output was 0.01%, almost entirely second harmonic. It was therefore presumably due to the IC, as any due to the lamp should be odd order. This is a little puzzling, as the op-amp specs at 0.003% THD typical. Nevertheless, it is quite a creditable performance for such a cheap, simple circuit. As an alternative, provision is made for the connection of an external modulation source, which naturally should not exceed 1Vpk-pk, or the modulation index will exceed 100% when R_{18} is at maximum.

Since it is easily incorporated and could come in useful, a DSBSC (double sideband suppressed carrier) mode was also included. In this position of switch S_1, the gain of IC_{7D} is doubled to give a bipolar drive at the modulator's Y_1 modulation input. R_{16} in conjunction with R_{22} permits zeroing of any offset at the output of IC_{7D}, and R_{23} was then adjusted for maximum carrier suppression in the DSBSC output.

Tests on the circuit, still in breadboard form, were carried out, with the following results. Figure 19.3(a) shows the maximum CW output at

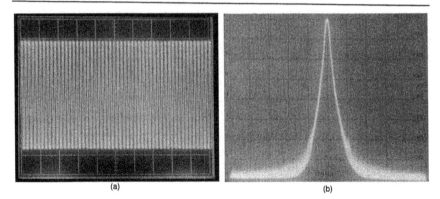

Figure 19.3 *(a) CW output at 10MHz into a 50Ω load (viz. a spectrum analyser), 'scope settings 0.2V/div. 0.5μs/div. (b) Spectrum of (a); centre frequency 10MHz, 20kHz/div. 3kHz IF bandwidth, video filter off, 10dB/div. ref. level (top of screen) 0dBm.*

10MHz while 3(b) shows the same signal displayed on a spectrum analyser. This indicates a level of 0dBm into 50Ω downstream of the source resistor R_{10}. Of the two, this is the answer I believe: the 'scope indicating about 1.1V pk-pk or over 4dB more than this. But you can't believe an X10 probe, with its 4 inch earth lead, even at as 'low' a frequency as 10MHz. In fact, Figure 19.3(a) is included solely for purposes of comparison with the AM case, as follows. Figure 19.4(a) shows the same 10MHz output but this time with 100% AM, with the same 'scope settings as before. It can be seen that the peak to peak voltage has increased slightly, and this is confirmed by the spectrum analyser picture, Figure 19.4(b). With 100% modulation, had the peak voltage been the same as before, the carrier component in Figure 19.4(b) would have been exactly −6dBm, i.e. 6dB below that in Figure 19.3(b). The fact that it is barely 5dB down indicates that while the levelling loop tries to control the peak level of the output, it is partly also sensitive to the mean RF level. If the time constant $C_{15}(R_9 + R_{12} + R_{35})$ were made short compared with the period of the modulation waveform, then mean level control would result and the carrier level would remain unchanged as the modulation depth were varied. But this would limit the highest usable modulating frequency to many octaves below the lowest usable radio frequency, at least in AM mode, and a restriction-free range for both modulation- and carrier-frequency operation was particularly sought. So the present scheme with (near) peak level control was retained.

Figure 19.4(b) shows that at 100% modulation, the second and third harmonic sidebands are only about 26dB and 40dB down respectively on the wanted fundamental sidebands. The second harmonic modulation is severe enough to be noticeable in Figure 19.4(a), as a reduction in ampli-

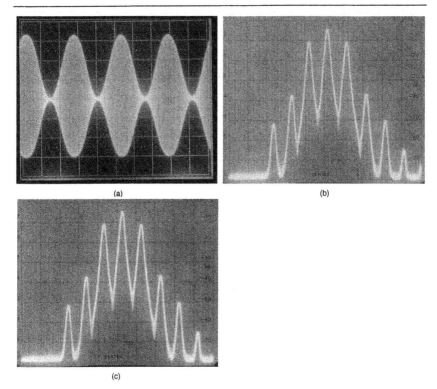

Figure 19.4 *(a) Output at 10MHz with 100% AM at 20kHz, 'scope settings 0.2V/div. 20μs/div. (b) Spectrum of (a); centre frequency 10MHz, 20kHz/div. 3kHz IF bandwidth, video filter off, 10dB/div. ref. level (top of screen) 0dBm. (c) As (b), but the input RF level reduced by 6dB. Switching to the 1dB division display indicated that the loop compressed a 6dB reduction in input level to a 0.2dB reduction in output level.*

tude of the negative-going peaks of the envelope relative to the positive. As it is the negative peaks which are sensed by the detector circuit, this largely explains the deviation from true peak-level control exhibited by the levelling loop, mentioned above. Figure 19.4(c) indicates just how effective the levelling loop is, compressing a 6dB reduction in input level to a 0.2dB reduction at the output. The great similarity between Figures 19.4(b) and (c) shows that the second- and third-order distortion sidebands arise not in IC_1 and not (according to its spec. sheet) in IC_2, but in IC_3 and IC_4, both of which are running just a few dBs below their 1dB compression point. A modification to the detector circuit to sense the positive peak, or better still a peak-to-peak detector, would be advantageous. In all modes – CW, AM and DSBSC – R_{29} provides the function of a 0 to 10dB output attenuator,

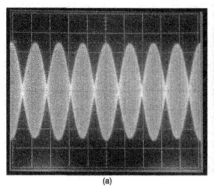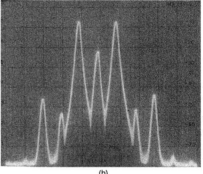

(a)   (b)

Figure 19.5   *(a) Output at 200MHz in DSBSC mode with 20kHz modulation. 0.2V/div vertical (but effectively uncalibrated at this frequency, on account of the probe earth lead inductance), 20µs/div. horizontal. (b) Spectrum of (a). Spectrum analyser settings as Figure 19.3(b) except centre frequency is 200MHz.*

although for the reasons indicated, when using AM the modulation should be adjusted for the desired depth before the output level is set.

Figure 19.5(a) shows operation at 200MHz in the DSBSC mode. Here, the Y_1 input of the modulator is taken both positive and negative, on alternate half cycles of the 20kHz modulating frequency. In consequence, the phase of the RF reverses twice per cycle of the modulation. Figure 19.5(b) shows the corresponding spectrum, from which it can be seen that the carrier is only some 15dB down on the 199.980MHz and 200.020MHz sidebands. This is despite the adjustment of R_{23} for maximum carrier suppression described earlier. The residual carrier is a component in quadrature with that controlled by R_{23}, and hence not affected by the nulling procedure. It is presumably due to capacitive feedthrough in, or around, IC_2. As is to be expected in this mode, the second-order distortion sidebands are way down, much lower than the third order. Note that in DSBSC mode, the output level is set solely by R_{29}, the 'output attenuator'. Any external modulation input should be set to 1V pk-pk and modulation depth control R_{18} to maximum. This is because, in DSBSC mode, 'modulation depth' is meaningless; whatever the modulation input level, the loop will always try to set the peak output level to that demanded by R_{29}.

The circuit shown will operate with input carrier frequencies down to about 1MHz; for operation down to lower frequencies all the capacitors in the RF path, such as C_1, C_3, etc., should be increased in value. Similarly, the external modulation facilities, whilst not DC coupled, should work down to a few hertz. The highest usable modulation frequency is set by the frequency responses of IC_{7C} and d, and IC_8; the response of the modulator's Y input (like its X input) extends up to 500MHz. Figure 19.6 shows a

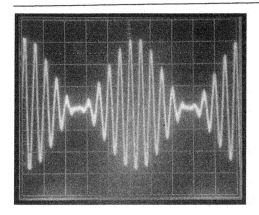

Figure 19.6 *10MHz output with 100% amplitude modulation at 1MHz, measured upstream of the 51Ω output resistor R_{10}. 0.2V/div. vertical, 0.2μs/div. horizontal.*

10MHz carrier 100% amplitude modulated at 1MHz, monitored at the junction of C_{15} and R_{10}. The modulation envelope seems visibly a very respectable sinewave, which surprised me greatly. That the LM13600 should work quite happily at this frequency was no surprise, its typical 3dB bandwidth is 2MHz. The TL084, on the other hand, will typically swing only ±2.5V into 2K at 1MHz, even on ±15V rails. So how was IC_{7C} coping on ±12V, with a load of around 500Ω? A quick check with a 'scope at IC_7 pin 8 showed that its output – a sinewave swinging entirely positive from 0V upwards – was distinctly poor. The positive peak was nicely rounded but the negative peak at 0V was distinctly pointed, although this doesn't show up very well in Figure 19.6. You can't get a quart out of a pint pot after all!

For my current requirements, the arrangements shown are adequate, and only remain to be recast in a tidier form than the present breadboard. But clearly the usable modulating frequency range could be greatly extended by substituting more modern, faster op-amps in place of IC_{7C} and d, if a faster variable gain amplifier were also used. To go with current feedback op-amps, the ultimate choice for the variable gain amplifier is obviously yet another AD834: you could then produce AM or DSBSC with modulating frequencies up to hundreds of megahertz if you felt so inclined.

20 Tweaking the diode detector

In the quest for a wide dynamic range detector, a number of variations on the 'infinite impedance' detector were tried. But in the end, a version of the straightforward diode detector, plus some additional linearisation circuitry, provided the best performance.

There are a number of important applications for RF detectors of wide dynamic range, for instance in various schemes for the linearisation of radio transmitters. The simple diode detector can be used, but suffers from a limited dynamic range and, a point which can be important, the fact that although its amplitude response in the large signal range is *linear*, it is *not proportional to the signal amplitude*. This is a result of V_f, the diode's forward voltage.

Figure A(a) shows an ideal diode which has an infinite slope resistance when the voltage at its anode is negative with respect to the cathode, and a slope resistance of zero when forward biased. Such diodes don't exist, but they can be closely approximated, over a limited frequency range, by a combination of one or more real diodes and op-amps. Figure A(b) shows a slightly less unrealistic representation of a real diode: the slope resistance when forward biased is still zero, but an anode voltage positive by an amount V_f relative to the cathode must be applied to cause current to flow. Figure 20.1(c) is one step nearer reality, showing as it does a finite diode slope resistance when forward biased. Unfortunately, the resistance does not drop instantly from infinity at forward voltages below V_f to a low constant value at V_f and above, but makes the transition gradually, as the curve at the bottom of the characteristic in Figure 20.1(d) shows. Projecting the characteristic in the forward-biased regime back until it cuts the voltage axis gives a value which may be taken approximately as the diode's V_f, the voltage which must be applied before some arbitrary small current flows.

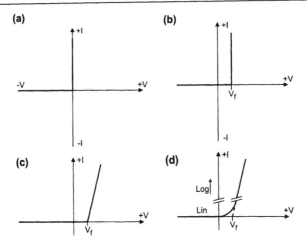

Figure A *(a) Current/voltage characteristic of an ideal (non-existent) diode. (b) Diode which is ideal except for forward voltage V_f which has to be overcome before current flows. (c) As (b) except that when conducting the diode has a finite resistance. (d) In a real diode, there is no sharp change of slope at V_f.*

When a practical Figure A(d) diode is used as an amplitude detector, the detected DC voltage output increases with RF input up to the point where the peak reverse voltage equals V_{br}, the diode reverse breakdown voltage. With decreasing RF input, the detected voltage falls linearly at first, then reaching a point where it falls faster than the input, since the latter no longer comfortably exceeds V_f. This is the 'square-law region', where the detected voltage is proportional to the applied power rather than to the applied voltage. Where it is simply desired to detect the presence of a signal, rather than accurately to measure its amplitude, the diode can still be successfully used in the square-law region, down to the level where the detected voltage starts to disappear into the circuit noise. This level of RF input to the detector is known as the tangential sensitivity, and can be determined by displaying the detected voltage on an oscilloscope whilst driving the diode with an RF signal 100% on/off modulated by a squarewave. The tangential sensitivity is the RF level at which the level of the top of the 'grass' in the off periods just concides with the level of the bottom of the grass in the on periods, as in the middle illustration in Figure B(a). The result obtained depends to some extent upon the oscilloscope's intensity setting and upon the operator, but being a quick, simple and easy test, it is widely employed. The tangential sensitivity can be improved by applying a small forward bias current to the detector diode as shown in Figure B(b) for various diodes. The optimum value of bias and the improvement that results depends

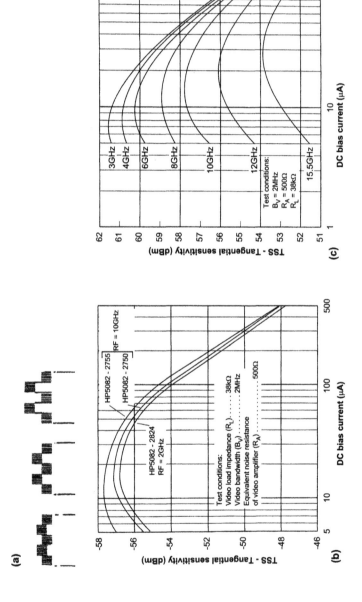

Figure B (a) Illustrating the tangential sensitivity (TSS) of a diode. (b) The tangential sensitivity can be improved by applying a small forward bias current to the detector diode as shown here for various diodes. (c) The optimum value of bias and the improvement that results depends upon frequency, as illustrated here for a particular diode. ((b) and (c) reproduced by courtesy of Hewlett-Packard Ltd, from HP Application Note 923.)

upon frequency, as illustrated in Figure B(c) for a particular diode. It can be seen that the tangential sensitivity can exceed -60dBm, so that the sensitivity of a simple diode-video receiver is only 35 to 40dB less than that of a superheterdyne receiver. Not only is the diode-video receiver much cheaper, simpler and easier to maintain, but it can easily be designed to offer a greater RF bandwidth than a superhet, and it is useful in a variety of applications, both military and civil.

However, as my interest at the time was in level measurement rather than just detection of the presence of a signal, thoughts turned to that useful circuit, the infinite impedance detector.[1] In this, the detector (a transistor, or in earlier times, a valve) is in its active range at all times, so one might think that the troublesome curve at the foot of the characteristic in Figure A(d) could be avoided. The plan was to surround the infinite impedance detector with a high gain servo loop which would jack up the DC voltage at the emitter of the detector, to restore the no-signal DC conditions. The transistor could be a type with a very high frequency cut-off f_t, and both the emitter and the collector could be bypassed at RF, confining RF solely to the base/emitter circuit. Identical DC conditions, regardless of whether RF was present or not, could be ensured by the use of constant-current generators in the emitter and collector circuits. (When two near-perfect constant-current generators – one a source and the other a sink – fight each other, their junction is a point of very high voltage gain.) The result was the pipedream circuit of Figure 20.1(a), where the op-amp supplies the difference between emitter's constant tail current I_e and the constant collector current I_c. The total gain within the loop includes a contribution equal to the slope resistance of the collector circuit divided by $(R_1$ plus $r_e)$, where r_e is the common base input resistance of the transistor. Some quick mental arithmetic showed this to be so great that the roll-off of loop gain had to start at such a low frequency that the circuit's response to a change of input level was inordinately slow. So the op-amp was replaced by a PNP transistor, IC_1 being redeployed to provide a voltage source for its emitter, Figure 20.1(b). Whilst this circuit worked, it represented little advance in low level sensitivity, in fact better results were obtained by removing the collector decoupling and closing the loop around Tr_1 and Tr_2 at RF as in Figure 20.1(c), which is similar to a circuit which appeared in Ref 2. Whilst this offered a useful improvement in low level sensitivity compared to a simple diode detector, the impression was left that all these circuit variations were basically more or less complicated ways of extracting such DC output as was obtainable from a basic diode detector circuit.

My attention was therefore redirected to the shunt diode detector circuit of Figure 20.2(a), which conveniently has one end of the diode grounded. The diode DC restores the input RF negative going with respect to ground,

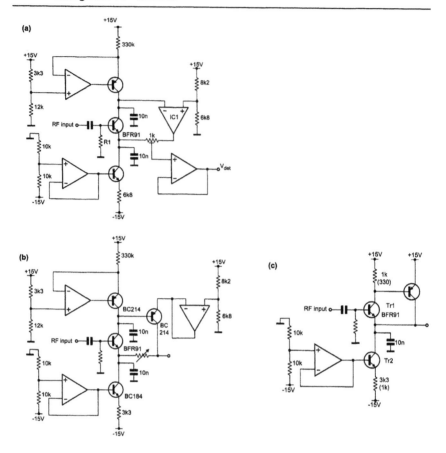

Figure 20.1 *(a) Infinite impedance detector embedded within a servo loop maintaining its collector current constant. (b) Variant of (a) with a lower (more manageable) amount of loop gain. (c) Variant of (b) where both transistors in the servo loop operate at RF.*

the smoothing circuit then picking out the mean level of the waveform at the diode's anode. The detected DC is thus equal to the peak value of the RF, or would be with an ideal diode where V_f is zero. The DC detected output of this circuit versus input RF level is the same as the usual diode peak detector circuit of Figure 20.2(b). The curve is linear at high levels but does not pass through the origin when projected backwards – i.e. the detected DC output is not proportional to the RF level. It could be made so by adding a constant offset equal to V_f to the detector's output. This would simply have the effect of raising the whole curve up by V_f, as in Figure 20.2(c). Although the output is now (at least at higher input levels) not only linear but also proportional to the input, there is a standing DC output V_f

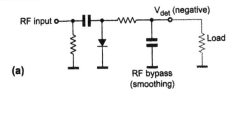

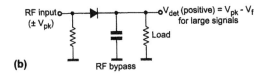

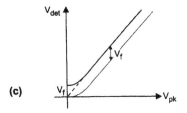

Figure 20.2 *(a) The shunt diode detector needs a DC path to ground at its output to enable it to follow a decreasing RF input level. It produces the same DC detected output for any input level as (b) the conventional peak detector circuit. (c) By adding V_f to the detected output, it becomes proportional to the signal level, except in the low level square-law area.*

when no RF is applied, an unfortunate state of affairs as V_f is temperature dependent. But with a little lateral thinking it is possible to add the V_f offset to the detected output in such a way that, as the detected output falls to zero, so does the added offset.

After a few iterations, a circuit designed to do just this finished up with the comparatively simple arrangement of Figure 20.3(a). This was built up on copper strip-board, except for the diode D_1, plus R_1, R_2, C_1 and C_2, which were all mounted at the back of a BNC socket, the body of which was soldered to a sheet of copper-clad board. To increase the sensitivity of the detector diode D_1, a small amount of forward bias is applied via a 10M resistor R_1, and the resultant offset balanced out by the corresponding drop across D_2, the 2M2 resistor being returned to the positive or negative 15V rail as necessary, depending on whether D_1 or D_2 exhibits the larger forward voltage at the current defined by R_1 and R_6. These diode voltages will of course be temperature dependent, but provided they track, they represent purely a common mode signal, and IC_{1a} is so connected as to

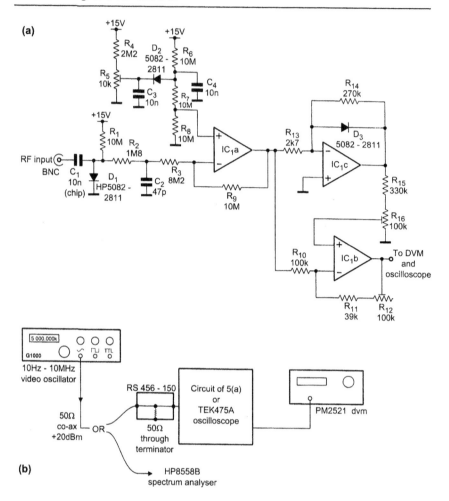

Figure 20.3 *(a) A detector circuit of extended dynamic range, with a DC output that is proportional to the level of the RF input. (b) Test set-up used to evaluate the circuit of (a). The 10dB steps of the G1000 generator were cross checked with an HP 355D VHF attenuator.*

reject any common mode input. R_2 with C_2 provides the RF smoothing to extract the detected DC component, while R_2 plus R_3 matches the value of the three other bridge resistors R_7, R_8 and R_9. IC_{1a} exhibits an inverting gain of unity to detected output of D_1, and its positive-going output is applied to inverting amplifier IC_{1b}, whose gain can be set to a gain of unity, or more or less as required. With the wiper of R_{16} set to ground, the detector law is exactly as the circuit of Figure 20.2(a).

The positive-going output of IC_{1a} is also applied to inverting amplifier

IC_{1c}. With R_5 correctly set up, the outputs of IC_{1b} and IC_{1c} will both be zero in the absence of any RF input, at least assuming all the op-amps ideal. With an increasing level of RF applied, the output of IC_{1a} will initially be small since it operates in the square-law region, but the output of IC_{1c} will rise much faster as it operates at a gain of 40dB, defined by R_{14}/R_{13}. However, with increasing detector output D_3 starts to conduct, progressively reducing the gain of IC_{1c}. Ultimately, the gain of IC_{1c} falls below unity, indeed almost to zero; in fact to ratio of the forward resistance of D_3 in the milliamps range (about 20Ω) to the value of R_3, or roughly −40dB.

By advancing the wiper of R_{16} from ground, a proportion of this voltage can be injected into the NI (non-inverting) input of IC_{1b}. (Note that to an input at its NI input, the gain of IC_{1b} is around 6dB, due to R_{10} and $R_{11} + R_{12}$.) This permits a useful degree of dynamic range extension, by ensuring that the output of the circuit as a whole starts to rise appreciably at a much lower RF input level than would otherwise be the case. The operation of the circuit was tested at RF levels of +20dBm downwards using the set-up in Figure 20.3(b), and the results recorded, Table 20.1.

Originally it was planned to use the video oscillator's maximum output frequency of 10MHz, but the spectrum analyser showed that there was significant second harmonic distortion at this frequency – bad enough to be clearly visible even on the oscillosope. The tests were therefore carried out at 5MHz, at which frequency the second harmonic was over 35dB down and all other harmonics much lower still. Harmonic distortion (especially even-order distortion) is an important consideration here, as the diode detectors of Figures 20.2(a) and (b) sense the peak of the RF waveform, although the indicated result is conventionally presented as the rms value of the signal, assuming this to be a pure sinewave. The 5MHz output of the generator was set to +20dBm using the indication on the spectrum analyser, after setting the gain of the latter using its internal CAL OUTPUT. As a cross check, the oscilloscope (recently calibrated by the manufacturer) was used to measure the peak to peak voltage at the output of the 50Ω through-load, i.e.

Table 1 *Measured performance of test set-up Fig. 4(b).*

Input dBm 50Ω	mVpk	Output Ideal normalised	mVpk	Actual normalised	Error %a
+20	3180	1.0	3180	1.0	−
+10	1005	0.316	946	0.297	−6
0	318	0.1	299	0.094	−6
−10	101	0.0316	85	0.027	−15
−20	31.8	0.01	31	0.0097	−3
−30	10.1	0.0032	10	0.031	−1.9
−40	3.2	0.0001	1.3	0.00041	−59

at the input to the detector circuit. The 'scopes answer was 6.4V pk-pk, perhaps a shade under; comforting agreement with the expected value of 6.36Vpk-pk for +20dB relative to 1mW in 50Ω.

Before taking the results recorded in Table 20.1, the circuit was set up as follows. With the wiper of R_{16} at ground and R_{12} set to mid-travel, R_5 was adjusted for zero reading on the DVM. The oscilloscope was also connected to the output of IC_{1c} to check that there was negligible hum pick-up, in view of the high circuit gain and the division of the circuit between two different boards. A 5MHz +20dBm input was then applied to the 50Ω through-termination and R_{12} adjusted to give the expected theoretical output of 3180mV. The input level was then reduced to −20dBm and R_{16} adjusted to give an output of 31.8mV. Results were not repeatable, due to the need for frequent resetting of R_5. So D_2 was relocated from the circuit board to a position adjacent to D_1, but well decoupled to prevent it picking up any RF. The rear of the BNC socket, with the two diodes and related components, was then enclosed in a box, shielding the diodes from both draughts and light, largely curing the drift problem, although a more modern quad op-amp with improved DC characteristics, in place of the TL084 used, would be even better. Following final adjustment of R_5, the +20dBm and −20dBm adjustments were then repeated, iterating them alternately until no further adjustment was needed. Later, a set of measurements were taken, the results shown in Table 20.1 being recorded. These were plotted on log/log graph paper over a 50dB range, Figure 20.4.

The results obtained to date are, as can be seen, very encouraging, though certain aspects await further investigation, notably the large negative error at around −10dBm. It must be said that the three diodes (HP5082 2811, Schottky barrier diodes for general-purpose applications) were not a matched set − being used simply because a number of them were in stock − although they are in fact available as matched quads (unconnected) under the type number 5082 2815. How much improvement this would achieve is an unknown quantity, as are a number of other possibilities. These include setting up at, say, +20 and −10dBm, instead of +20 and −20dBm, or choosing a different value for the forward bias applied to D_1 and D_2. Other parameters available are varying the ratio of R_{14} to R_{13}, and also their absolute values, and the inclusion of resistance in series with D_3. These might help to reduce the effect of a mismatch between two separate characteristics. The detector law is a square law at low levels, whilst the compensation diode law, determining the voltage across D_3, is basically a log law relative to the voltage applied to R_{13}.

Where maximum sensitivity is required from the circuit, one of the low 1/f (flicker) noise diodes from the 5082 2xxx series might be a better choice, the 30V V_{br} of the 5082 2301 (for example) permitting the detector to accept inputs up to almost +34dBm against only +27dBm for the 5082 2811. For the widest possible dynamic range, the 5082 2800 (1N5711) with

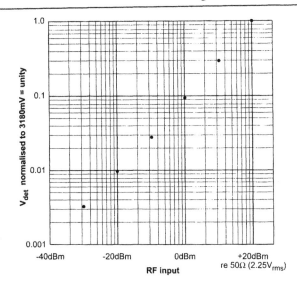

Figure 20.4 *The results from Table 20.1 over a 50dB range, plotted on log/log graph paper. The vertical axis is the actual detected DC output voltage and the horizontal indicates the peak input voltage, the measured points being plotted as crosses. The sloping line at 45° indicates the output that would be provided by a perfect detector. For convenience in plotting, both the peak input voltage and the DC voltage have been normalised to 3.18V equals unity.*

its V_{br} of 70V would accept an input of up to +40dBm. Assuming that it too would work down to −30dbm in the circuit of Figure 20.5(a), it would provide a detector with a dynamic range of over 70dB.

Detectors of wide dynamic range can be used in the linearisation of transmitters using a non-constant envelope type of modulation, e.g. AM, DSBSC and all varieties of SSB from compatible AM, through pilot carrier types right through to pure J3E. An NFB loop from the detected audio recovered from a coupler at the transmitter's output, feeding back to the exciter, can suppress third-order intermodulation products in a solid state transmitter to 60dB below PEP (peak envelope power),[2] provided the transmitter is designed to hold AM to PM conversion in all stages to a low level. This can provide a much cheaper solution than complex arrangements such as a polar loop or cartesian loop scheme.

References

1. Hickman, I. 'Measuring detectors', *Electronics World and Wireless World*, Nov. and Dec. 1991.
2. UK Patent 2209639A, Single Sideband Transmitters, 1991.

21 Oscillating at UHF

Two basic circuit configurations are responsible for most oscillator designs working at frequencies up to the UHF range. This article describes the often conflicting requirements of UHF oscillators – including a disadvantage of the emitter follower now put to good use.

Oscillators for frequencies to UHF and beyond have been built using all sorts of active devices, from valves onwards. The majority use three-terminal active devices, often connected to a simple tuned circuit in one of two basic ways. These were enumerated for my benefit as a student, by an older colleague of many years' experience, with the aid of a sketch which I call 'O'Connor's Universal Oscillator Circuit' and which is reproduced as Figure 21.1. The circuit is drawn in an unconventional way to emphasise the points which follow. For the circuit to function as an oscillator, Z_2 and Z_3 must be reactances of the same sign – both inductances or both capacitances – while Z_1 must be of the opposite sign. With this proviso, the figure shows that relative to the cathode (emitter, source), the voltages at the other two electrodes are in antiphase. No earth connection is shown, since in principle the circuit could be provided with the necessary power supplies via ideal RF chokes of infinite reactance at the operating fre-

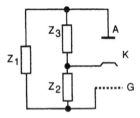

Figure 21.1 *O'Connor's universal oscillator circuit. Z_1 is a reactance of one sign whilst Z_2 and Z_3 are both of the other. For valve read NPN bipolar, N-channel FET, HEMT, etc., as appropriate.*

quency, and a, g or k earthed as convenient, or the whole circuit left floating. If Z_1 is an inductor with capacitors at Z_2, Z_3, the circuit is a Colpitts oscillator, whilst if a tapped inductor forms Z_2 and Z_3 with Z_1 being a capacitor then the circuit is a Hartley oscillator. One way or another, all three electrodes of the active device must be connected to the tuned circuit.

Many other circuit arrangements are possible, some using more than one active device, a variety being shown in Figure 21.2. However, at UHF a circuit using a single device, connected as in Figure 21.1, often proves best. This is because additional phase shifts associated with a second active device or parasitics associated with coupled windings introduce additional complexities into the design process, effects that would be smaller or negligible at VHF or HF.

As a basis of a signal generator, an oscillator with a wide tuning range is required. Whilst at one time this would have been tuned by a precision mechanical variable capacitor, in a more modern application varactor tuning will usually be employed, permitting accurate frequency control by means of a PLL – phase lock loop. With a possible application in view, I experimented with what might be regarded as a Colpitts oscillator – if you draw in the transistor's internal base/emitter capacitance to go with the 3.3pF external collector/emitter capacitance as Z_2 and Z_3. The circuit is shown in Figure 21.3(a). The problem with a wide-range oscillator is that one wants to be able to vary its frequency over a wide range at will, but then instantly have its frequency as stable as a rock once one has set it to any particular desired frequency. As a start therefore, it pays at the outset to design the oscillator circuit to have very stable DC conditions, ensured in Figure 21.3(a) by the supply regulator, and by the base bias chain with its low source resistance at DC, which is moreover well decoupled at RF. As first constructed, the oscillator covered from under 400MHz to over 600Mhz, but was modified as shown for the intended purpose to cover well in excess of 200MHz centred on 400MHz. This is shown in Figure 21.3(b), where the oscillator was tuned back and forth across its range during the 6 second exposure required by the home-made oscilloscope camera (which does duty also for my spectrum analyser). There is a general slope in level of several dB across the tuning range, but the superimposed ripples are due to the connection to the spectrum analyser. This was demonstrated by doubling the length of coax used for the connection, resulting in twice as many ripples. Clearly the analyser's input impedance isn't exactly 50Ω on the most sensitive range used; switching in 10dB at the input attenuator largely removed the ripples.

It is a convenient fiction that in common cathode/emitter/source mode, an active device is an inverting amplifier, i.e. that the voltages at the other two electrodes are in antiphase. This is true in the case of valves up to quite high frequencies, since the velocity of electrons *in vacuo* is a good deal faster

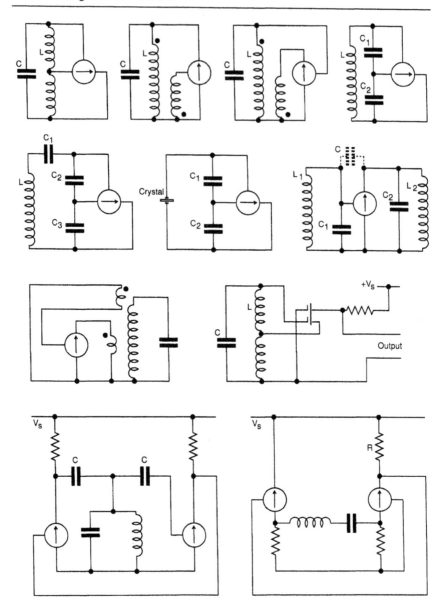

Figure 21.2　*A variety of oscillator circuits, some more suited to lower frequencies.*

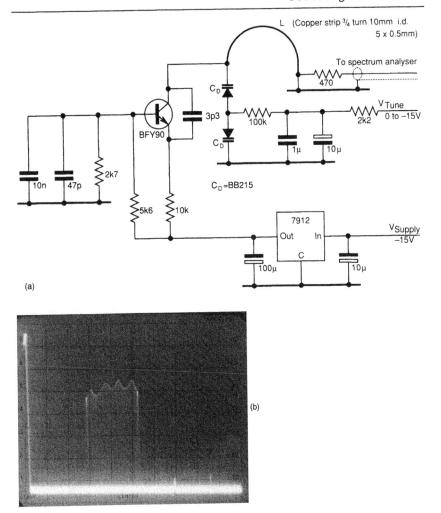

(a)

(b)

Figure 21.3 *(a) A wide-range UHF varicap-tuned oscillator circuit. (b) The tuning range covers from below 300MHz to over 500MHz. Span 0–1000MHz, 10dB/div. vertical.*

than minority carriers in silicon. But in a transistor, phase shifts start to show up even in high frequency devices, at quite a low frequency. This is illustrated in Figure 21.4. Figure 21.4(a) shows the relation between the currents in the three electrodes of a transistor at DC, and recaps on the relation between the current gains α and β – the latter is often also called α' or h_{FE}. Figure 21.4(b) shows how even quite a small phase shift in the collector current can result in a phase shift in the base current which is much larger, and moreover in the opposite direction. (In the simplified

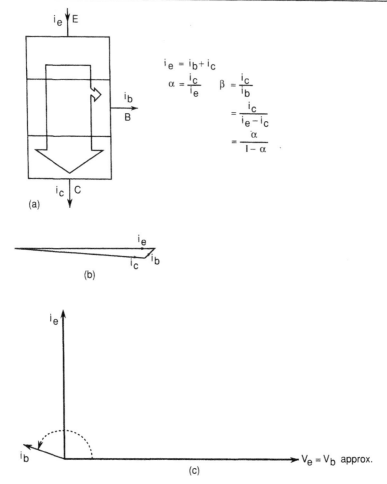

$$i_e = i_b + i_c$$

$$\alpha = \frac{i_c}{i_e} \qquad \beta = \frac{i_c}{i_b}$$

$$= \frac{i_c}{i_e - i_c}$$

$$= \frac{\alpha}{1 - \alpha}$$

(a)

(b)

(c)

Figure 21.4 *(a) Showing the relation at DC (0Hz) between the current in the electrodes of a transistor, the common base gain α, and the common emitter gain β (or α'). (b) Showing how even quite a small phase lag in the collector current (with respect to the emitter current) results in a much larger phase advance in the base current. (c) If the load on an emitter follower is capacitive, so that the emitter current leads the emitter voltage, the base current will lead the base voltage by an angle well in excess of 90°, resulting in a negative resistance component at the input.*

treatment given here, any phase shifts suffered by the base or collector currents after they part company – due to 'transmission line delay' in different regions of the bulk of the semiconductor – are assumed to be negligible.) The higher the DC value of β (i.e. the more nearly the magnitude of the collector current equals that of the emitter), the smaller the phase shift in the collector current needed to give a 45° advance to the base current. For an audio frequency transistor such as the BC109 with its typical β of 300 and f_T of 300MHz, this occurs at around 1MHz. At higher frequencies, the base current can lead the emitter current by not far off 90°.

The emitter follower is an extremely useful and widely used circuit, acting as a buffer and permitting a high impedance source to drive a lower impedance load. But the circuit has an unfortunate tendency to oscillate – particularly if the load is a bit capacitive, the phase advance suffered by the base current being the culprit. This is illustrated in Figure 21.4(c), where an important assumption is made: that the mutual conductance of the device is high (its output impedance low compared to the impedance of the load connected to the emitter) so that to a first approximation the voltage at the emitter equals that at the base. In Figure 21.4(c) (where it is assumed an emitter follower is driving a purely capacitive load), the emitter current will be leading the base voltage by up to virtually 90°. And with the base current substantially leading the emitter current, it follows that the base current leads the base voltage by well over 90° – the input impedance consists of a negative resistive component in parallel with a capacitance. This effect has been used as the basis of a microwave oscillator design producing over 100mW output at 2GHz.[1] It can equally well be used at UHF, and Figure 21.5 shows just such an application. The reactance of 18pF at 345MHz is 25Ω (doubtless effectively reduced somewhat by the inductance of the leads, even though these were kept as short as possible), so the emitter circuit load is almost purely capacitive. The capacitance tuning the inductor consisted only of the capacitive component of device input impedance, and device and circuit strays.

If the circuit of Figure 21.5(a) is compared with that of Figure 21.3, it will be seen to be almost identical. In both cases, the collector is connected to the opposite end of the tuned circuit from the base, whilst a capacitor is connected from the emitter to the collector end of the tuned circuit. Thus in fact most oscillators operating at VHF or above and using a single active device are likely to be found on analysis to be negative resistance oscillators. Depending on the Q of the tuned circuit (and that in Figure 21.5(a) was certainly not very high), the noise performance or short-term stability of such an oscillator can be good, though of course the medium- and long-term stability will be poor unless the oscillator is used as a VCO (voltage controlled oscillator) in a phase lock loop. Figure 21.5(b) shows the output of the Figure 21.5(a) circuit, the centre frequency being 345MHz and the dispersion (horizontal scale) 5kHz/div. The analyser bandwidth was set to 1kHz and a great many sweeps occurred during the 6 second exposure needed to record the background and graticule. Some noise modulation is evident but the overall shape is not so very different from

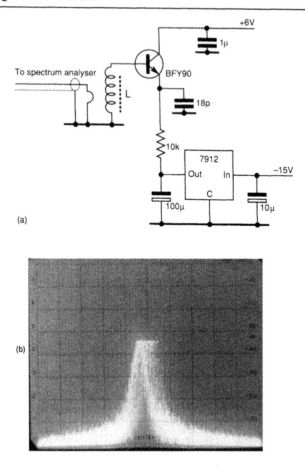

Figure 21.5   *(a) A UHF oscillator using the negative input resistance effect. The tuning capacitance consists of the capacitive component of base circuit input impedance plus device and circuit strays. L consisted of 3 turns (spaced one wire width) of 16SWG TCW, 5mm i.d. with 3.75mm ferrite slug. (b) Output from the loosely coupled 1 turn winding, centre frequency 345MHz, 5kHz/div. horizontal, 10dB/div. vertical, ref. level –10dBm, IF bandwidth 1kHz, video filter off.*

that of the analyser's 1kHz filter. However, it can be seen that towards the end of the exposure, the oscillator took it into its head to start wandering up in frequency – a stability of 1kHz in an open loop UHF oscillator could be achieved, but only with a more sophisticated circuit, using a high Q cavity resonator, for example.

Another arrangement providing improved frequency stability without resorting to a PLL (phase locked loop) is the line stabilised oscillator. The

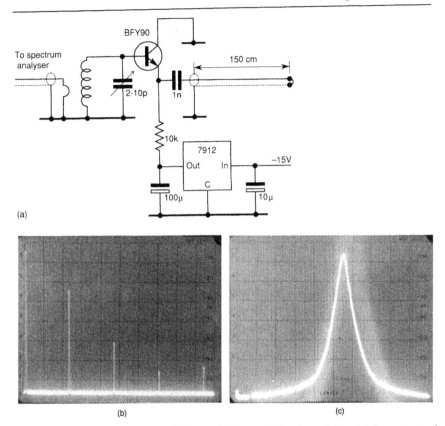

(a)

(b)   (c)

Figure 21.6 *(a) A simple and fairly crude line-stabilised oscillator. (b) The output of (a), span 0–1000MHz, 10dB/div. vertical, ref. level (top of screen) –10dBm. (c) The fundamental component of (b), centre frequency 235MHz, dispersion 5kHz/div., IF bandwidth 1kHz, video filter at max. (giving a post detector bandwidth of 1.5Hz), 10dB/div. vertical, ref. level –30dBm.*

circuit of Figure 21.5(a) was modified to work in this mode, the line consisting of 150cm of 50Ω miniature coax, believed to have a velocity ratio of around 0.66, with its far end shorted. A tuning capacitor was added to enable the tank circuit to be tuned to a frequency at which the emitter load looked capacitive. The circuit oscillated at 235MHz, at which frequency the length of the line would be just over one and three quarters wavelengths, i.e. capacitive.

Clearly there will be other frequencies, both higher and lower, at which the line looked capacitive, for example where the line length is 5/4λ, 9/4λ, 11/4λ, etc., and the tuned circuit is used to pick out one of these as the operating frequency. If the tank circuit Q is high and the regeneration only just sufficient to ensure oscillation, then only one of these modes can be sustained. If the tank Q is lower and the negative resistance much lower than necessary to sustain oscillation, the circuit can oscillate in several modes at once. This was the case when the collector supply was the same as in Figure 21.5(a). Oscillation in several modes simultaneously was prevented by the simple expedient of reducing the collector voltage until it equalled the base voltage, as shown in Figure 21.6(a). With a constant tail current generator or RF choke plus resistor combination in place of the 10K resistor to −12V, clearly the oscillator circuit would happily work on a supply of a volt or two. The output from the loosely coupled winding was as in Figure 21.6(b), where the span is 0–1000MHz. The fundamental at 235MHz is visible, together with the second, third and fourth harmonics. Figure 21.6(c) zooms in onto the fundamental, at 5kHz/div. horizontal. At the selected video filter bandwidth, a single sweep took 6 seconds, and at 60dB down, the response is 15kHz wide, which is more or less identical to the analyser's 1kHz filter spec.

Of course, a length of coax does not make for a very convenient line stabilised oscillator, and even if semi-rigid solid outer coax were used, the stability of the oscillator with temperature would not be wonderful. But line stabilisaton is now very attractive and competitive, in the form of SAW (surface acoustic wave) resonators. Owing to the extremely slow propagation speed of acoustic waves in lithium niobate (slow at least compared with the speed of light), a compact package can contain a line length of many wavelengths. Such devices are used at UHF in lieu of crystals, where tight frequency control is required. An example is the range of 418MHz telemetry modules featured in Ref. 2.

Connecting a negative resistance across a tuned circuit results in an oscillator, and the negative resistance need not imply a three terminal device. Many years ago a two terminal device – the tunnel diode – was a popular means of making UHF oscillators, at a time when transistors with adequate performance were not available, or at best very expensive. Now that transistors with more than adequate performance are common and cheap, the tunnel diode UHF oscillator has taken a back seat, but negative resistance two-terminal oscillator circuits are still used at microwave frequencies, in the form of the Gunn diode oscillator.

Acknowledgments

Figure 21.2 reproduced from the *Newnes Practical RF Handbook* by permission of the publishers, Butterworth-Heinemann.

References

1. Partha and Krishnakumar, 'Oscillator design employs common-collector bipolars', *Microwaves and RF*, Oct. 1994 pp. 88–92.
2. Hickman, I. 'Low power radio links', *Electronics World and Wireless World*, Feb. 1993 pp. 140–144.

22   Modulating linearly

Due to crowding in the radio spectrum at HF, it is important to reduce third-order (and higher odd-order) intermodulation products to a minimum. This article looks at a means of achieving a highly linear modulator – useful as a test adjunct in testing HF PA stages, and for other purposes. The reasons behind the need for low distortion modulators are spelt out more fully in a reply to a reader's letter, reproduced at the end of the article.

Receiving a given radio signal may be commercially important or it may even be a matter of life and death. But the ether is a crowded place, and a weak signal – even though it be in an otherwise unoccupied channel – may be drowned by the spillover of energy from a higher powered transmission in an adjacent channel. This is particularly the case on the crowded HF bands, and there have been persistent rumours of an impending tightening of the specifications for the level of transmitter third-order intermodulation products. Since I drafted, some years ago now, an amendment to these (incorporated in the current issue of CCIR Recommendation 326), time has passed but the rumours persist.

In an HF SSB transmitter, it is likely to be the TX PA (transmitter power amplifier) output stage that is principally responsible for third-order (and possibly even higher odd-order) intermodulation products. But problems can also arise in the modulator stage, especially if poorly designed, whilst for test and measurement purposes as clean a test signal as possible is desirable.This reminded me of an article published many years ago, concerning the linearisation of an active double-balanced modulator, using op-amps. A search through my file of articles published in a particular magazine proved fruitless, so I suppose that memory must have played me false and the article is in one of my other files, which cover over a dozen different electronics magazines. So I am unable to quote the reference at the moment, but the scheme is certainly attractive and worth another look.

Since it was the principle of the scheme that was of interest, I decided to investigate it at (very!) low frequencies, which would enable the investigation to be undertaken without the need for careful and elaborate construction to avoid problems with parasitics. It would also give an opportunity to use the latest addition to my collection of test equipment. To provide a two-tone test signal to the modulator, two video oscillators were used, one set to 1kHz and the other to 1.2kHz. Each was separately connected to a 5Hz–50kHz spectrum analyser and its spectrum stored, both traces being shown in Figure 22.1(a). The two tones just reach up to the reference level (top of screen) which is at +10dBV. At the left of the display, adjacent to the 0Hz marker, low levels of 50, 150 and 250Hz can be seen, being odd multiples of the mains frequency and doubtless due to stray field from a mains transformer. Also visible are 100Hz sidebands either side of each tone, at about 70dB down. (Unfortunately the scale does not show up

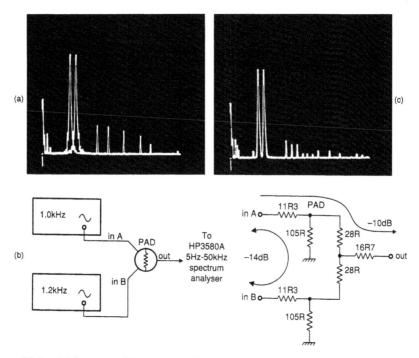

Figure 22.1 *(a) Spectra of the two oscillators used to provide the two-tone test signal. Span 0–5kHz, resolution bandwidth 10Hz, post detector smoothing off, refer. level (top of screen) +10dBV, noise floor – 90dBref. (b) 50Ω resistive combining pad providing 10dB attenuation from each input to the output and 14dB isolation between inputs when terminated in 50Ω. (Nearest E24 values used.) (c) Using the combining pad (unterminated), a third-order intermod is visible at 800Hz, as well as sum and difference products. Spectrum analyser settings as (a).*

because the HP3580A does not feature graticule illumination, and I have not yet added this facility to the camera.) The second harmonic of each tone at 2kHz and 2.4kHz is also visible at over 60dB down, while higher harmonics are lower still.

When testing a modulator, it is obviously essential that the two-tone test signal is itself free of intermodulation products. This is not so straightforward as it sounds, requiring isolation between the outputs. The 50Ω outputs of the two video oscillators were therefore combined using a special resistive pad designed for this purpose, Figure 22.1(b). It is basically a three-port 6dB resistive combiner with an extra 4dB pad in series with two of the ports. Thus the attenuation from each input to the output is 10dB, whilst the isolation (attenuation) between inputs is 14dB. The two tones were combined using this pad and the output connected to the spectrum analyser, Figure 22.1(c). The two tones are now 10dB down on the reference level, due to the attenuation of the pad, the spectrum analyser's sensitivity being unchanged, and all the other products are similarly 10dB lower. It is of course not permissible to increase the analyser's sensitivity to set the two tones back to the reference level, as their combined value when in phase (their 'PEP' or peak envelope power) is 6dB greater than either tone alone; enough to cause intermodulation products within the analyser itself, due to overload.

The isolation between the oscillators is evidently insufficient, as a number of products are visible. These include the difference frequency, visible as an additional 200Hz line at the left of the trace between the 150Hz and 250Hz lines of Figure 22.1(a), and the sum frequency at 2200Hz, between the two second harmonics. More importantly, there is a third-order intermodulation product at 200Hz below the 1kHz tone. Third-order intermods usually come in pairs, but there is only the barest hint of a product at 200Hz above the 1.2kHz tone. The reason may be that the output of each oscillator, feeding back into the output circuit of the other, is producing upper and lower third-order intermods, with the phasing such that they add on the lower side but cancel on the upper.

In search of a better arrangement, the special combining pad was removed and the output level of each oscillator reduced by 10dB to compensate. The two tones were then added at a virtual earth point, as in Figure 22.2(a). This resulted in the two tones appearing at the same level as in Figure 22.1(c), but with no third-order intermods visible above the noise floor of 80dB below either tone, 86dB below PEP, see Figure 22.2(b).

With a clean two-tone test signal available, it was time to look at the performance of an active double-balanced modulator. The one chosen was the popular and widely second-sourced LM1496, the internal circuit of which is as in Figure 22.3(a). The standing current through each of the signal input transistors Q_5 and Q_6 is set by the associated current sources Q_7 and Q_8. The standing currents are modified by the signal current

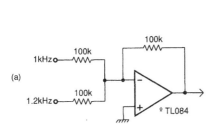

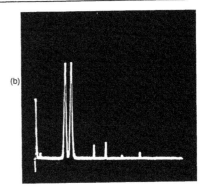

Figure 22.2 *(a) Combining the two tones at a virtual earth point provides near perfect isolation between sources – (b) – as the spectrum analyser shows. Settings as Figure 22.1(a). Generator outputs reduced to ca. +7dBm in 50Ω, to give 0dBv unloaded.*

through the gain-defining resistor, which is connected between pins 2 and 3. The current through this resistor will of course be zero when when the differential signal voltage between pins 1 and 4 is zero. The standing currents through Q_7 and Q_8 mirror the current injected into Q_9 via the bias terminal. The current through Q_5 is steered by the switching cell Q_1–Q_4 to the + output or the − output at the same time as that through Q_6 is steered to the − output or the + output, by the carrier input. Thus if the currents in Q_5 and Q_6 are equal, the current in the + output is independent of the instantaneous polarity of the carrier, and similarly for the − output: i.e. both outputs are balanced as far as the carrier is concerned. With signal present, when the current through Q_5 increases and that through Q_6 decreases, RF current appears at the + output in phase with the carrier, or in antiphase when the current through Q_6 exceeds that through Q_5, the situation at the − output being the exact reverse. However, the baseband signal itself does not appear at either output – the circuit is thus balanced also as far as the signal is concerned, or 'double balanced'.

The mixer was connected up as in Figure 22.3(b), which is generally similar to the published typical application circuit, but using ±12V supplies instead of +12V and −8V. A 20kHz squarewave of ±1.5V peak was applied to the carrier input. Since the switching stages Q_1 and Q_2, Q_3 and Q_4 have no emitter degeneration resistors, a small swing of 100mV or so is enough to switch the current from one path to the other. The 10% to 90% rise and fall times of the 3V pk-pk squarewave carrier were just 1.5µs, so the effective switching time was less than 100ns, or very small indeed compared to the period of the carrier. A high degree of linearity at the signal port is ensured by the comparatively large value of resistance

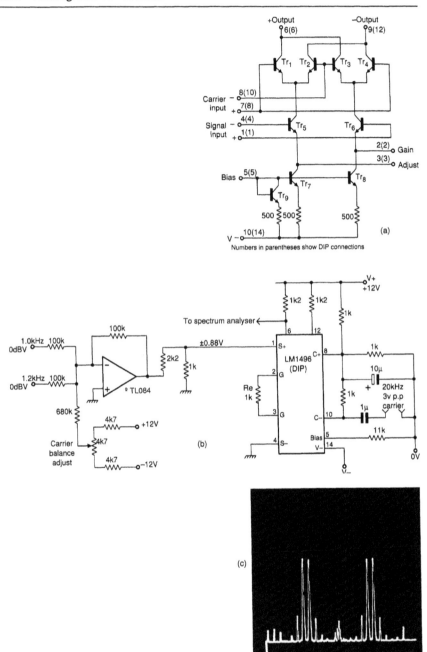

6(6)

9(12)

Tr_1 Tr_2 Tr_3 Tr_4

Carrier − 8(10)
input + 7(8)

Signal − 4(4)
input + 1(1)

Tr_5 Tr_6

2(2) Gain
3(3) Adjust

Bias 5(5)

Tr_7 Tr_8

Tr_9

500 500   500

V − 10(14)

(a)

Numbers in parentheses show DIP connections

V+
+12V

1k2 1k2

1k

To spectrum analyser ←

6   12

100k

1.0kHz 100k
0dBV

±0.88V

1   S+
LM1496
C+ 8

1k

0.1μ

2k2
1k

1.2kHz 100k
0dBV

2   G

10μ

20kHz
3v p.p
carrier

TL084

Re
1k

1k

680k

3   G

1μ

C− 10

4k7
+12V

11k

Carrier
balance
adjust

4k7

4k7
−12V

(b)

4   S−

Bias 5

V− 14

0V

V−

(c)

Figure 22.3

between pins 2 and 3, namely 1kΩ. Very much greater sensitivity to the input at the signal port can be obtained by reducing the value of this resistor, but only at the expense of reduced linearity, as discussed in the Box. The signal was applied to pin 1 only, pin 4 being grounded, i.e. as an unbalanced input. But the effect is almost exactly the same as using a balanced drive, in view of the near-perfect current-source 'long tails' supplying Q_5 and Q_6 emitters. Similarly, an unbalanced drive was employed for the carrier input.

Whatever the value of the gain-defining resistor between pins 2 and 3, as the voltage swing at the signal input modulates the two tail currents nearer and nearer to ±100% of their standing value, S-shaped or third-order distortion will eventually set in. This will cause third harmonic distortion of the signal current fed to the switching section $Q_1 - Q_4$. This will appear not only as third harmonic distortion of the two tones, but also as third-order intermod products either side of the two tones, in exactly the same way as in an audio frequency amplifier. These products are at frequencies $2.f_1-f_2$ and $2.f_2-f_1$, in the present case 800Hz and 1400Hz. However, rather than appearing at baseband as in an audio amplifier, here the intermods are translated along with the two tones to the upper and lower sideband outputs of the mixer.

One of the mixer circuit outputs was applied to the input of the spectrum analyser, as in Figure 22.3(b), the resultant spectrum being as in Figure 22.3(c). It shows the (largely) suppressed carrier at centre screen, the two tones in the upper sideband with third-order intermods either side, and a similar picture in the lower sideband. The two tone inputs to the mixer were −10dBV each, and the third-order intermod products were 50dB down on PEP (44B down on either tone). Fifth-order products at 600Hz and 1600Hz above the carrier (and below it) are also just visible. With 1mA tail current in each of Q_5 and Q_6, the maximum possible linear current modulation is ±1mA, via the gaindefining resistor R_e. This corresponds approximately to 2V pk-pk at the signal input. With two tones at −10dBv, the peak envelope voltage is ±0.88V, or barely 1dB below that theoretical maximum!

Figure 22.3 *(a) The LM1496 double-balanced modulator internal circuit diagram. (Reproduced by courtesy of National Semiconductor.) (b) The LM1496 connected as a double-balanced modulator, with a two-tone test signal applied, modulating the 1mA standing current in Q_5 and Q_6 by ±90%. (c) The resultant output spectrum, showing the (largely) suppressed carrier at centre screen, the two tones in the upper sideband with third-order intermods either side, and a similar picture in the lower sideband. The third-order intermod products at 20.8kHz and 21.4kHz are 44dB down on either tone, i.e. 50dB down on PEP (and similarly in the lower sideband). Centre frequency 20kHz, span 5kHz, resolution bandwidth 10Hz, post detector smoothing off.*

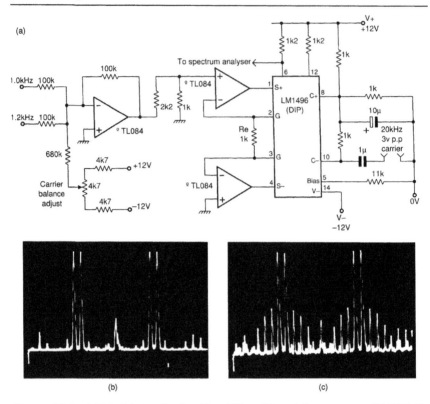

Figure 22.4   *(a) Modulator circuit with additional linearising op-amps. (b) With the same input signal level as in Figure 22.3, the third-order intermods are now about 65dB down on P.E.P. and the fifth order barely discernible. Spectrum analyser settings as in Figure 22.3(c). (c) With both tones increased in level by 3dB, the signal input is overloaded by about 2dB, and a mass of higher order products appears.*

Clearly then, the high value of gain-defining resistor is very effective at linearising the modulation. A reduction in the drive level, so that the signal was not just 1dB below overload level, but 3dB, 6dB or more below, would improve the linearity, driving the intermods even further below PEP. But the noise level remains unchanged, so the circuit's dynamic range would be reduced. However, a substantial reduction in intermod levels is possible without reducing the input signal level at all, by using op-amps to linearise the transconductance of the Q_5, Q_6 pair. The arrangement is simplicity itself, Figure 22.4(a). Connected as unity gain followers, the two op-amps drive the bases of Q_5 and Q_6 so as to force the voltages at their emitters to equal the signal input at the op-amps' non-inverting inputs. The fall in differential transconductance of the Q_5, Q_6 pair as one or other nears cut-off is within the loop feedback, and thus largely overcome. This is shown in Figure 22.4b), where the analyser settings are as those in Figure 22.3(c). Now, the third-order intermods are 65dB down on PEP, even with the input within 1dB of overload.

With both tones increased by 3dB, the signal input circuit is overdriven and a mass of odd-order intermod products of higher orders appear, Figure 22.4(c). Thus the linearisation allows the circuit to operate in an extremely linear manner, right up to just below the theoretical overload point. And although, for convenience, the results presented here were obtained when operating the LM1496 at very low frequencies, the scheme doubtless operates at much higher frequencies. The data book typical performance curves for the LM1496 show its performance to 50MHz and beyond, and op-amps are now available with gain-bandwidth products of many hundreds of MHz. Thus the linearisation scheme should be readily implemented at, say, 10.7MHz, or even higher frequencies, providing a superlinear modulator for test and measurement purposes, or for use in transmitters of exceptional linearity, using perhaps class A output stages.

Box

The long-tailed pair Q_5, Q_6 in Figure 22.3 operates very linearly, provided the signal input swing is not too large, due to the emitter degeneration provided by the gain setting resistor R_e. But the transistors themselves contribute some additional resistance, and this is dependent on the emitter current. When a grounded emitter transistor is driven from a very high impedance source (a constant current generator), the collector current is determined principally by the base current and the device's current gain. But when driven from a very low impedance source (a constant voltage generator), a different model is appropriate. Often, a very simple model, such as shown at (a) in Figure 22.5, suffices to give an understanding of how a circuit works, and of its limitations. The figure shows a transistor with infinite mutual conductance g_m, so that as far as small changes of signal voltage are concerned, its internal emitter voltage follows exactly the base voltage. But between this notional internal emitter and the outside world, there is a resistance r_e, the value of which depends upon the emitter current and hence also on the signal voltage. The resistance r_e is inversely proportional to the emitter current, in fact at room temperature the value of the resistance is given by

$$r_e = 25/I_e \, \Omega$$

where I_e is in milliamps.

Now imagine that the transistor is biased so that the DC (standing) current is 1mA. Then when a ±250µV AC signal is applied to the base, the current swing will be ±10µA almost exactly – but not quite. For

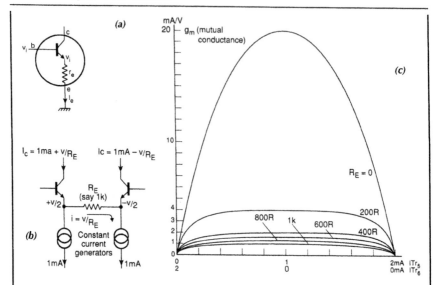

Figure 22.5 *(a) Simple model of a transistor driven from a constant voltage source. (b) In a long-tailed pair with ideal 'tails', the increase in current through one transistor must exactly match the decrease through the other. (c) Additional resistance R_e between the emitters linearises the stage, permitting a much larger percentage current swing for a given acceptable degree of distortion, at the expense of needing a much larger voltage swing at the input (reduced stage gain).*

when the emitter current rises to 1.01mA, r_e will fall to $25/1.01\Omega$, and likewise will rise by 1% when the current falls to 0.99mA. So the increase in current at the positive peak will be slightly greater than the decrease at the negative peak. The disparity becomes greater as the fractional modulation of the emitter current is increased, leading to significant second harmonic distortion unless some measure, such as negative feedback, is used to control it.

In a balanced circuit, such as (b) in the figure, the internal emitter resistance r_e of each transistor must be added to any external resistance R_e connected between the transistors. With constant tail currents as shown, as the current through one transistor increases, that through the other must fall by the same amount. If the current through Q_5 decreases by 10% its r_e will rise to $25/0.9 = 27.78\Omega$ while the r_e of Q_6 will fall to 22.72Ω. Thus the effective emitter-to-emitter resistance is $(R_e + 50.5)\Omega$. This differs from $(R_e + 50)\Omega$, the value when both emitter currents are 1mA, by only 1% even if $R = 0$, and a mere 0.05% if $R = 1k\Omega$. If the current variation is not $\pm1\%$ but, say, $\pm90\%$, then even with $R = 1k\Omega$, significant peak crushing (third harmonic distortion) will result, due to the variation of the effective emitter–emitter resistance.

> For when the current through one transistor doubles to 2mA, its r_e drops only to 12.5Ω, whereas for the other, when I_e falls to 0mA, r_e rises to infinity. The effect of different values of emitter-emitter resistance R_e in linearising the mutual conductance g_m of a long-tailed pair is illustrated in (c) in the figure.

Clarified linear modulation

The Editor

Dear Sir

In his letter concerning my article 'Modulating linearly' (July 1995 edition), a reader makes the point that '. . . intermodulation products are usually generated at the power amplifier final stage . .'. I entirely agree, indeed the second paragraph of the article runs 'In an HF SSB transmitter, it is likely to be the transmitter power amplifier output stage that is principally responsible for . . . intermodulation products.' It goes on to point out that as clean a test signal as possible is desirable for test and measurement purposes.

Nevertheless, it is true that HF SSB PAs only produce the amount of intermodulation products commonly observed, because they are permitted to do so by current regulations -25dB below either tone for R3E, J2E and H3E without privacy device; 35dB with privacy device and for A3E, B8E, R7B, B7B and B7W, per CCIR Recommendation 326. There is no incentive for manufacturers to produce 'cleaner' PAs bearing in mind that this would involve extra costs. However, if this were necessary, the required techniques are already to hand. References 1 and 2 refer to the Polar Loop technique, which was intended to permit the use of SSB with 5kHz channel spacing at VHF, should this standard ever be introduced. PA intermod products of 55dB below either tone were demonstrated, and the principle would be directly applicable at HF. The same sort of reduction in odd-order intermod levels could be achieved by the related cartesian loop system, which I believe was also developed at Bath University.

These techniques require the resolution of a sample of the TX output into its real and imaginary components at IF in order to close the loop. This naturally requires a fair amount of kit, so in the absence of mandatory regulations requiring that sort of performance, it is not surprising that extensive use has not been made of these schemes. In the mid-eighties, I developed a simpler arrangement, which was applied to a 150W broadband HF PA module. The latter was designed to be multicoupled up in stages, to provide various powers up to 1kW PEP.[3] This scheme also reduced the third-order intermod products to 60dB or more below PEP.

Even with this degree of transmitter output stage linearity, it remains true that the main use for an ultralinear modulator is in test and measurement. For any out of band intermod products produced by the modulator will usually be adequately suppressed by the following sideband filter. But there is an important exception: namely where modulation is performed without a sideband filter, e.g. by quadrature modulation.

However, I cannot agree that odd-order intermodulation products add to the intelligibility of speech *per se*. Indeed they degrade it. The improvement in HF communications in difficult conditions arises from the fact that, with a very poor signal-to-noise ratio, the degradation due to intermods is more than offset by the improved signal-to-noise ratio brought about by the increased average radiated power. The instantaneous compression provided by IF clipping followed by a second sideband filter (compared with the slower compression provided by a VOGAD) is very effective at emphasising the quieter components of speech such as unvoiced consonants – particularly sibilants – which otherwise get lost in the noise.

Yours faithfully
Ian Hickman

References

1. Petrovic, V and Gosling, W. 'Polar loop transmitter', *Electronics Letters* 1979, 15(10), pp. 286–288.
2. Petrovic, V and Smith, C N. 'The design of VHF SSB transmitters', IEE Conference on Communications Equipment and Systems, April 1982 pp. 150–155.
3. 'Single sideband transmitters', UK Patent 2209639B granted 1 May 1991, assigned to Siemens Plessey Electronic Systems Ltd.

23 Sweeping to VHF

A sweeper is a useful piece of kit in any RF development laboratory. But once commonplace, they are now found only incorporated into a spectrum or network analyser, as a tracking generator. Spanning 0–200MHz in one range, this stand-alone sweeper provides an output level of +3dBm ±1dB into 50Ω, with very low harmonic and spurious content.

Sweepers (or swept frequency generators) were once commonplace items of test equipment in electronics laboratories. The earliest types, known as 'wobbulators', used a vibrating capacitor and consequently produced an output whose frequency varied approximately sinusoidally with time. A vibrating capacitor was also used in certain wartime radar altimeters, which were therefore much sought after on the post-war Government Surplus market by would-be constructors of wobbulators. Later models provided an output frequency which varied linearly with time, a great convenience when displaying the response of, for example a receiver on an oscilloscope. Nowadays, stand-alone sweepers have been relegated to history, but a sweeper is incorporated in most modern spectrum and network analysers, its output frequency being at all times that to which the receive section of the instrument is tuned.

The construction of a spectrum or network analyser is a substantial task for anyone interested in electronics to undertake, but a sweeper is a much simpler undertaking, and as I hope this article will show, the resultant instrument is useful in a number of ways. The instrument described covers 0–200MHz, or any desired narrower sweep within this range, providing an output level of +3dBm ±1dB into 50Ω, with very low harmonic and spurious content. An external attenuator may be used to reduce the level as required. Sweep speeds of 25ms for the full 200MHz sweep (or such smaller sweep as may be set) down to 250s are provided. The faster speeds are suitable for use with any simple 'scope, whilst the slower speeds are provided for use with a DSO (digital storage oscilloscope). Additionally, the sweep may be turned off. The unit may then be used to provide a CW output, or alternatively an external audio input up to 15kHz may be connected to provide an FM output. The maximum deviation available is the full 200MHz, and as the external input is DC coupled, the unit may be used as a 0–200MHz VCO in a phase locked loop.

A sweeper is basically a beat frequency oscillator, BFO, consisting of a fixed frequency oscillator (FFO), a mixer and a voltage-controlled oscillator or VCO. The latter can be swept from the same frequency as the FFO to some higher maximum frequency, the outputs of both oscillators being connected to the mixer. The mixer output, suitably low-pass filtered and amplified, forms the sweeper output. In addition to covering the full range 0Hz to $f_{(VCOmax)}-f_{(FFO)}$, arrangements are always included to permit the output to sweep as small a range as desired, centred anywhere within the full range.

Building a sweeper is greatly simplified if extensive use is made of RF ICs, rather than doing it all with discretes. The associated low frequency circuitry, sweep generator, linearisation, etc., will of course use ICs anyway – the universal quad op-amp being the obvious choice. The writer was prompted to develop this instrument for his own use by the appearance of the Analog Devices AD831 low distortion mixer. Being an active mixer, it has the advantage that the conversion gain can be set to any desired figure, the usual value being 0dB, compared with the typical 6.5dB conversion loss of a passive Schottky diode ring mixer. The data sheet for this active mixer, housed in a 20 lead PLCC, claims a 500MHz bandwidth for both the RF and LO ports. Some of the graphs showing typical performance extend to 600MHz, and knowing that the larger, more reputable IC manufacturers tend to give conservative figures in their data sheets, it was decided to aim at a design providing an output frequency range covering 0–200MHz. For reasons that will appear later, this implies an FFO at just over 400MHz and a VCO covering from there up to just over 600MHz.

As the RF department was clearly going to be the tricky part, design and construction of this was commenced first. Clearly, a fully screened construction was essential, so a diecast box was initially considered. But it was soon abandoned, since three separate internal compartments, completely screened from each other, would be required for the the two oscillators and the mixer/output department. These are not easily implemented in a diecast box, so a box was fabricated out of copper-clad SRBP. Single sided was used for the base, four sides and three lids, with double sided for the internal partitions, as indicated in Figure 23.1. Each lid sits just below the top of its compartment, supported on stops soldered to the compartment wall. With its copper side uppermost, when the RF unit is complete, each lid can be soldered to the walls of its compartment. (It is recommended to tack it at one central point only on each of its four sides, as you are sure to want to remove it again sometime. The box seams, by contrast, are

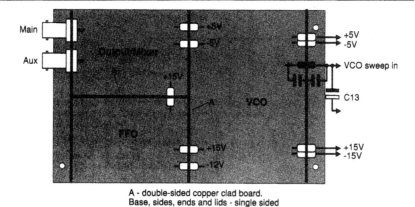

A - double-sided copper clad board.
Base, sides, ends and lids - single sided

Figure 23.1 *Construction of the screened box for the RF unit.*

soldered along their whole length.) For ease of access during construction, initially, just the base, sides and central partition were assembled. The short partition and ends were fitted only when the three sections were all basically functional, the lids only when final testing was complete.

This approach illustrates one of the differences between general analog circuit design and RF design which tends to be overlooked by the newcomer to RF. At RF, the mechanical design has to be considered before the detailed circuit design, though of course the designer must have a reasonably clear idea of what sort and how much circuitry the mechanics has to accommodate.

The VCO was designed first, accommodated in the largest of the three compartments. It, and the remaining RF circuitry, are shown in Figure 23.2. In the author's prototype, the VCO used the same inductor as the FFO, with the result that it covers the required tuning range, but with not much to spare at the top end, whilst the bottom end extended down to around 300MHz. Hence the different VCO inductor specification shown in the figure.

The output of each oscillator was taken from a tap near the earthy end of the inductor, to ensure low loading on the oscillator, thus not greatly reducing its operating Q (which – especially in the case of the VCO – is much lower than one would wish). The outputs are amplified to about −10dBm, and connected to the mixer via miniature 50Ω coax. Each coax is taken into the mixer compartment under a notch in the intervening partition. At this point, the outer sleeve was locally stripped, and the coax screen soldered to the partition. Similarly, to prevent leakage of RF energy from one compartment to another, power rails are taken through 10n feedthrough capacitors. These are also used to bring the supplies into the VCO end of the RF unit, but a different arrangement – shown in Figure 23.2 – is needed to bring in the VCO tuning voltage.

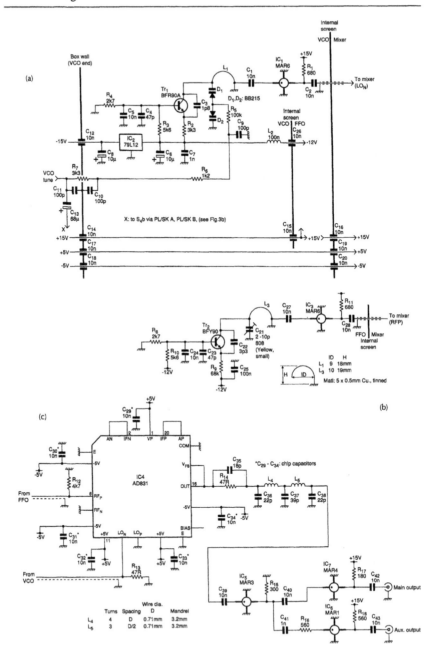

Figure 23.2   *The circuitry of the RF unit. (a) The VCO circuit and supplies decoupling. (b) The fixed frequency oscillator. (c) the mixer and output stages.*

The VCO drive is applied to the LO input. In the AD831, this input feeds an internal limiting amplifier before being applied to the mixer cell, resulting in the mixer performance being quite unaffected by variations in the level of LO drive as the VCO sweeps across the band. The RF port is a linear input, and thus the waveform of the mixer's IF output reflects that of the RF input. The FFO is designed to produce an output with low harmonic content and the following amplifier also operates way below the 1dB compression point. Thus the mixer's IF output also exhibits a low harmonic content, and the use of an MAR4, with its +11dBm, 1dB compression point, ensures that at the sweeper's modest +3dBm output, the harmonic content remains low.

It can be seen that extensive use is made of the very convenient and economically priced 'Mini Circuits' RF amplifier ICs, available from various distributors. In addition to the MAR4 providing the main +3dBm output, an MAR1 provides a −13dBm auxiliary output, well buffered from the main output, which may be connected to a digital frequency meter.

Compared to the RF department, the construction of the sweeper's low frequency circuitry is very non-critical; in the case of the prototype it was all accommodated on an RS stripboard 433-826, supported horizontally on brackets at the rear of the case, above the mains tranformer and the RF unit. Connections between this board and the RF unit were made via an 8-way 0.1inch pitch Molex plug and socket, mounted at the left rear of the stripboard, above the RF unit. In front of the 8-way connector, a 15-way Molex plug and socket provided all the necessary connections to the front panel controls and the BNC Ext FM input and TRIGGER output sockets, the whole instrument being mounted in a case $305 \times 159 \times 133$mm high, Maplin LH42V. The suite of stabilised supplies were mounted at the right-hand side of the stripboard, above the $20V + 20V$, 25VA mains transformer, Maplin DH28F.

Figure 23.3(a) shows the centre frequency setting circuitry IC_{9B}, with 10K coarse and fine controls, together with the VCO drive amplifier IC_{10}. A TLC2201 was chosen for IC_{10}, since its CMOS output stage can swing to within millivolts of either rail, enabling the VCO to be tuned up to a top frequency just in excess of 600MHz. For ease of adjustment, both the 10K pots were 10-turn types. A select on test resistor R_{24} was fitted at the earthy end of the coarse tuning control R_{23}, such that its tuning range (in CW mode, sweep disabled) was limited to just below 0Hz, in fact about −10MHz. Note that both IC_9 and IC_{10} operate between 0V and −15V rails, to ensure that the varactor diodes in the VCO can never become forward biased. IC_{10} incorporates a linearisation network around its feedback resistor.

Figure 23.3(b) shows the remaining control circuitry, principally the sweep generator IC_{8B} and IC_{8C} (operating on +15 and −15V rails), sweep width buffer IC_{8D} and level translator IC_{9D}. Note that the sweep generator

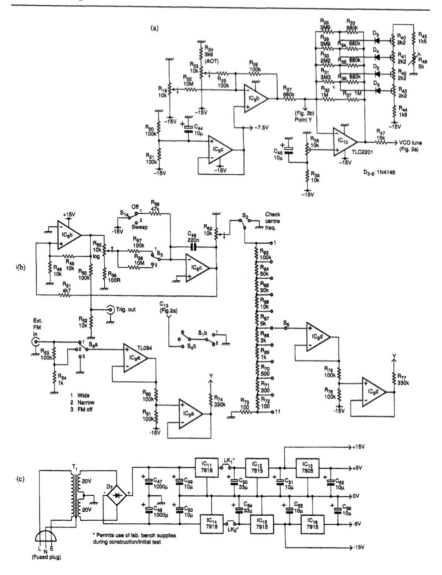

Figure 23.3 *The control and supplies board, etc. (a) Centre frequency setting and linearisation stages. (b) The sweep generator and EXT FM input stages. (c) The power supplies.*

produces a triangular voltage, not a sawtoooth. Note also that the highest VCO frequency results when the tuning voltage is at -15V and that IC_{10} is inverting. Consequently, when displaying, for example, the discriminator characteristic of an FM receiver on an oscilloscope, the oscilloscope should be triggered at the start of the positive-going half of the triangular waveform. A suitable time base speed should be selected to display the whole of the frequency sweep. With the values shown in Figure 3(b), the wiper of sweep speed variable control R_{55} at the top of its travel and sweep speed switch S_3 in position 1, the sweep time is about 25ms. Thus a sweep speed of 2ms/div., adjusted to 2.5/div. with the 'scopes time base speed VAR control, will prove suitable. (Alternatively, a front panel output socket may provide a suitably scaled version of the sweep voltage from IC_{8C}, for application to the 'scope's X input, in XY mode.) Progressively slower sweep speeds may be set by means of R_{55}, whilst position 2 of S_3 provides sweep speeds slower by a factor of 100, for use with DSOs in ROLL mode. A buffer and level translator are provided for the Ext FM input, IC_{8A} and IC_{9A}. Figure 23.3(c) shows the power supplies department, which uses 78/79XX series IC regulators, fitted with heatsinks.

The sweep generator produces a linear sweep voltage or ramp, but applied directly to the VCO this will not result in a linear frequency sweep. The sweep linearisation network around IC_{10} shapes the ramp into a non-linear form which is the inverse of the VCO's tuning voltage/frequency characteristic, resulting in a linear frequency sweep. The necessary adjustments to the linearisation are most conveniently made whilst displaying the full 0–200MHz sweep, with the aid of 10MHz 'birdie markers'. These can be produced by feeding the output of the sweeper to one input of a passive Schottky diode double-balanced mixer, and narrow 10MHz spikes to the other, the mixer output being displayed on an oscilloscope. This arrangement was tried, but proved unsuccessful, partly due to the poor rise time of the 10MHz squarewave which was obtained from a 10Hz–10MHz video generator. But even below 100MHz, where birdie markers were obtained, their amplitude was very small. Now, in principle, the linearisation network could be set up without the sweep, using the centre frequency setting control R_{23}, but in practice this proves impossibly cumbersome. So a special harmonic mixer was constructed, as in Figure 23.4. This used the 10MHz squarewave already mentioned, but with its edges sharpened up by a test circuit that I had designed previously.[1] I strongly recommend retaining any special test circuits you may build for possible reuse at a later date.

Figure 23.5(a) shows the output of the harmonic mixer (lower trace) during a positive-going varactor tuning voltage sweep, upper trace – this corresponds to a falling frequency sweep. The sweep extends from 10MHz to −80MHz, i.e. to the point where the VCO frequency is 80MHz below the FFO frequency. Notice the absence of a birdie marker at 0Hz. This photograph was taken before fitting any linearisation around IC_{10}, and the decreasing tuning sensitivity with increasing frequency is clearly visible.

In the design stage of the sweep circuitry it was felt that five breakpoints should be ample to achieve an acceptable degree of linearisation, and provision for these was therefore made in the circuit, Figure 23.3(a). When

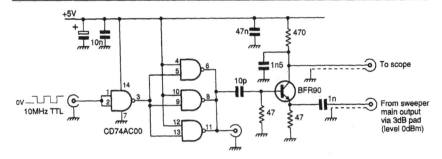

Figure 23.4 *Harmonic mixer used to provide birdie markers, to facilitate the adjustment of the frequency sweep linearisation.*

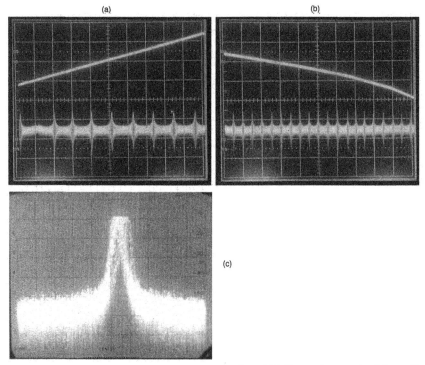

Figure 23.5 *(a) Sweep voltage (upper trace) and birdie markers output from the harmonic mixer (lower trace) as the frequency sweeps from 10MHz through zero to −80MHz. (b) Linearised sweep from 0 to 200MHz. (c) Output in CW mode, Ref. level +10dBm, 10dB/div. vertical, centre frequency 100MHz, 5kHz/div. horizontal, IF bandwidth 1kHz.*

it came to setting up the linearisation, never one to use two resistors where one will do, I tried the effect of using just three of the breakpoints. Adequate linearity was not achievable, but the addition of a fourth break-point as in Figure 23.3(a) achieved a good degree of linearity, Figure 23.5(b). This shows the VCO tuning voltage (top trace) and the output of the harmonic mixer, bottom trace, showing birdie markers at 10MHz intervals from 0Hz to 200MHz.

As viewed on the oscilloscope the fit is not quite perfect, though this does not show too clearly in the figure. However, as the trace is expanded from the 20MHz/div. shown, by means of coarse sweep width control S_5 (fine sweep width R_{62} being fully clockwise) to, say, 1MHz/div., any deviation from perfect linearity becomes more noticeable. It is therefore worthwhile taking the extra trouble to use all five breakpoints. R_{40-43} set the positions of the individual breakpoints, whilst R_{46} permits the adjustment of where the first beakpoint occurs. R_{28-31} set the strength of each breakpoint, and were 5MΩ pots during the setting up, being replaced with the nearest E12 values when linearity had been achieved. A final tweaking of the positions of the breakpoints was then carried out.

Owing to the non-linear gain of IC_{10} (following linearisation), if its non-invert input were returned to -7.5V, on applying a sweep voltage, its output would swing further in the negative-going direction than in the positive. To counteract this, it is returned to a somewhat lower voltage, set by R_{38}. Whatever sweep setting is selected, the sweep may be turned off momentarily by means of the check centre frequency switch S_2, a biased toggle. Owing, again, to the non-linearity of IC_{10}, the 'centre frequency' does not correspond to midscreen on the oscilloscope display, but to a point about two divisions to the right of this. As the sweep width is reduced, the display will expand off each side of the screen, points to the right of 'centre frequency' moving to the right, points to the left moving leftwards.

Switch S_1 provides a choice of Sweep On (position 2) or Sweep Off, and serves two purposes. In the Off position, C_{13} is switched into circuit, preventing any AC noise on the VCO Tune line reaching the varactor (provided Ext FM In is set to Off at S_4). This reduces the linewidth of the RF output substantially, Figure 23.5(c). Towards the end of the six seconds exposure required by the oscilloscope camera the frequency has started to wander up by a few kilohertz, but apart from this it can be seen that the line width is not greatly in excess of the analyser's 1kHz IF bandwidth (centre frequency 100MHz, 5kHz/div. horizontal, ref. level $+10$dBm, 10dB/div. vertical), although noise 'shoulders' appear at only some 40dB below the output. The broadband noise floor is about -70dB, compare Figure 23.5(c) with Figure 23.8(a), which covers a 1MHz span, as against a 50kHz span.

Note that to use CW mode, it is necessary to set fine display width R_{62} to minimum and S_5 to position 11. This is because in the Sweep Off position,

S_1 causes IC_{8C}'s output to rise to +15V, which would set the VCO to maximum frequency.

The other use for S_1 is when using a DSO in ROLL mode as the display. Setting S_1 to Sweep Off (i.e. CW mode) sets the VCO to maximum frequency and the Trig. Out level positive. The DSO may now be set to RUN in ROLL mode, with positive trigger. On returning S_1 to the sweep position, the output will sweep down to minimum frequency and then back up to maximum, at which point the trigger output will go positive again, halting the DSO acquisition. If a suitable sweep speed was selected with R_{55} and S_3, to match the DSO's time/div. in ROLL mode, the screen will display the result of one complete frequency sweep from minimum to maximum.

The linewidth or close-in noise of the instrument is much worse than that of a signal generator, but this is normally the case with sweepers, or any other BFO-type instrument. However, in other respects, the performance is very good. Figure 23.6(a) shows the output at 20MHz (span 0–200MHz, ref. level +10dBm, 10dB/div.) from which it can be seen that the second harmonic is over 40dB down and the third harmonic nearly 50dB down on the output, which is around +3dBm. Even at 100MHz, Figure 23.6(b), the second harmonic is nearly 40dB down, whilst higher harmonics are of course out of band. The level flatness of the main output is within ±1dB over 2–200MHz. These photographs were actually taken with the unit in CW mode, but the harmonic performance is of course identical in sweep mode.

Another important criterion of any signal generator, including sweepers, is the level of spurious outputs. These are an unfortunate fact of life, wherever two or more oscillators and a mixer are involved. They can be explained with the aid of Figure 23.6(c), which covers 0–500MHz, ref. level +10dBm, 10dB/div. vertically. The output at 180MHz is the main output at a level of +3dBm nominal. The output at 360MHz is its second harmonic (which is still considerable, despite being in the stopband of the post mixer low-pass filter). The three other outputs are spurious responses or 'spurii'. For purposes of explanation, assume the FFO is at 400MHz, so that its second harmonic is at 800MHz. To produce a 0–200MHz output, the VCO must cover 400–600MHz. As the VCO frequency approaches 600MHz, it will beat with the second harmonic of the FFO, this beat frequency falling to 200MHz as the VFO frequency reaches 600MHz. This is a 'two/one' response, i.e. twice the FFO frequency minus the VCO frequency. (The other two/one response, 2 × VCO–FFO is out of band at all times.) For this reason, the FFO frequency is set very slightly higher than 400MHz; in the prototype, measurement showed it to be 406.6MHz. (With the aid of a frequency meter and spectrum analyser, this is easily deduced from the measured frequency f_x of the wanted output at which it coincides with the two/one response. For then, VCO–FFO and

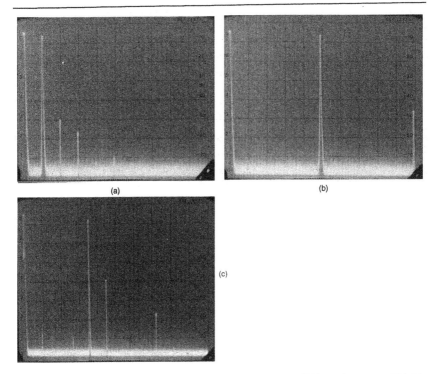

Figure 23.6 *(a) Showing the sweeper main output, in CW mode, set to 20MHz. Note the low level of the second and third harmonics. (Span 0–200MHz, ref. level +10dBm, 10dB/div. vertical.) (b) As (a), but output set to 98MHz. (c) Main output at 180MHz, showing spurious responses (see text). Settings as (a) except span 0–500MHz.*

$2 \times$ FFO–VCO both equal f_x.) This means that the two/one response never comes below 206.6MHz, i.e. it remains out of band at all times.

The two smallest outputs are higher order spurious responses; fortunately such responses get smaller the higher the order. The output at 133.4MHz is a three/four response, as can be deduced from the fact that it moves 30MHz for every 10MHz that the wanted output moves; so it must involve the third harmonic of the VCO. Given the main output is 180MHz, the third harmonic of the VCO will be $3(180 + 406.6) = 1759.8$MHz. The fourth harmonic of the FFO is 1626.4MHz, so the 133.4MHz output (which is over 60dB below the main output) is the $3 \times$ VCO–$4 \times$ FFO response. As the main output is tuned up or down, this spurious response moves in the same direction, eventually as the frequency is reduced, disappearing out of band below 0Hz, never to return.

The other spurious response, at 46.6MHz, moves at twice the rate of the main output, and in the opposite direction. Similar reasoning to the above identifies it as a two/three response, namely $3 \times$ FFO$-2 \times$ VCO. Since it travels in the opposite direction from the main output, their frequencies can coincide: this happens at a frequency of 135MHz. A spurious output that can occur very close to an expected response could be most embarassing, but fortunately its level is around 60dB down on the main output.

Another criterion of a good sweeper is how low a frequency it can be tuned to before the two oscillators begin to affect each other, eventually locking up to the same frequency. Figure 23.7(a) shows the main output set to 600kHz, ref. level $+10$dBm, 10dB/div. vertical, span 0–5MHz. Note the numerous harmonics, the second being only 10dB below the main output. This is due to the two oscillators affecting each other, due to inadequate isolation between them. One oscillator tends to keep in step with the other until the disturbance from its undisturbed phase becomes excessive, when it suddenly slips a cycle. This is shown in the time domain in Figure 23.7(b), where the frequency has been reduced still further to about 65kHz. At this frequency the power in the fundamental was greatly reduced, and a host of harmonics up to very high orders present. (Forgetting the frequency for the moment, electronic organ buffs will recognise the waveform as similar to that of a trumpet stop.) Any attempt to tune the two oscillators closer together simply resulted in their locking up to the same frequency.

The Figure 23.7 results were taken during the development of the RF unit, before all of the screening was fitted. In the finished instrument, lock-up did not occur until the offset was below 25kHz, or a mere 125 parts per million of the frequency of the two oscillators. This attests to the

TIME BASE = 5uS
CH1 V/DIV = 100mV

(a)   (b)

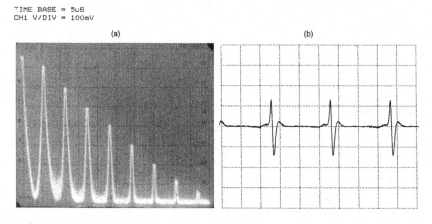

Figure 23.7 *(a) Main output of the sweeper in CW mode, set to 600kHz. Ref. level +10dBm, 10dB/div., span 0–5MHz. (b) Waveform of the main output of the sweeper in CW mode, set to 65kHz. The two oscillators run at the same frequency except when slipping a cycle every 16µs.*

efficacy of the screening arrangements, the RF to LO port isolation of the AD831 mixer and the reverse isolation of the MARx series amplifiers. As a result, the sweeper is quite usable down to 1MHz, even at 500kHz, the harmonics are still over 25dB down, though the output level has fallen to just over 0dBm, due to the size of the various coupling capacitors.

Figure 23.8(a) is an example of the sweeper output when used in the external FM input mode with the FM input switch set to Narrow. The centre frequency is 100MHz and the horizontal scale 100kHz /div. Ref. level and vertical scale are +10dBm and 10dB/div., as usual. It shows the output frequency modulated at 1.5kHz with a peak deviation of ±75kHz, the maximum deviation in the broadcast FM standard. Individual FM sidebands are visible, but they are blurred due to the lack of an exact relation between the modulating frequency, the maximum deviation and the sweep repetition frequency of the spectrum analyser, over the exposure time required by the 'scope camera.

Figure 23.8(b) is another example of the instrument in use, this time in sweep mode. The top trace shows the discriminator output of a 'tranny portable', an ITT KB Junior Super AM/FM model dating from the 1970s. The 'scope settings were 0.2V/div. and 2ms/div. The sweeper was set to sweep 3MHz during 20ms, and no direct connection between the set and the sweeper was necessary, stray pick-up sufficing. The resultant output is shown on the upper trace. The discriminator characteristic is about the width one would expect for a mono receiver, but the peaks are slightly asymmetrical and the central portion of the characteristic is not quite as

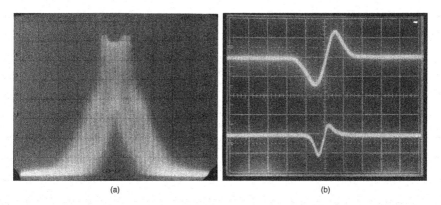

(a)   (b)

Figure 23.8 *(a) Main output of the sweeper in FM mode, showing modulation at 1.5kHz with a peak deviation of 75kHz. Ref. level +10dBm, 10dB/div., centre frequency 100MHz, 100kHz/div. (b) Output of the discriminator circuit of a 'tranny portable', upper trace (0.2V/div., 2ms/div.) with a suitable sweep rate. Distorted output, lower trace, when the sweep speed is excessive (0.2V/div., 200μs/div.)*

linear as it might be: clearly after so many years, the set is in need of realignment. The lower trace shows the effect of excessive sweep speed. The FM broadcast standard allows for a ±75kHz deviation at a maximum modulating frequency of 15kHz, and this corresponds to a maximum rate of change of frequency of 7139Hz/µs. In the lower trace of Figure 23.8(b), the 3MHz sweep width has been increased to 75MHz, and the 'scope time base speed to 200µs/div. The rate of change of frequency is now over 93000Hz/µs, or over ten times the maximum specified rate. The result is that the discriminator output is grossly distorted and reduced in amplitude. This demonstrates that when using a sweeper, the sweep rate selected must be appropriate to the bandwidth of the circuit under test.

There are many other uses for a sweeper (used with a return loss bridge, for example, it performs much the same job as a scalar network analyser), and it is hoped to explore some of these in a future article.

Reference

1. See Hickman, I. 'CFBOs, delivering speed at any gain?', *Electronics World and Wireless World*, Jan. 1993 pp. 78–80, reproduced in the *Analog Circuits Cookbook*, Butterworth-Heinemann 1995, ISBN 0 7506 2002 1.

Part 4
BASIC PRINCIPLES

24 Charting RF performance

> The Smith Chart has been around for decades, and provides a neat way of representing RF impedances – including those observed at device input and output ports. Understanding the chart is important, since it is one of the common formats (and arguably the most useful) in which impedance measurements are presented on a network analyser.

In RF circuits, achieving optimum performance usually depends upon obtaining a good match between a source and a load, one of which is often a resistive impedance of 50Ω (I am ignoring for the moment those cases where a degree of mismatch is deliberately arranged to achieve stability in an amplifier circuit). For instance, the input circuit of an IC designed for RF applications may present an input impedance looking like a lossy capacitance. A network of two or more reactances (just occasionally, one will do) may be used to bring it to a 50Ω resistive impedance, at least at the desired operating frequency and possibly over a rather wider bandwidth. Numerous articles covering the design of 'L' and more complicated networks have appeared over the years and the necessary calculations may often be expedited by expressing an impedance in terms of a series resistance and reactance, or as a parallel combination of conductance and susceptance. Converting from one to the other is straightforward, if cumbersome, see Box. Also involved is the addition of impedances or admittances, and this is easily achieved if they are in $R + jX$ (resistance and reactance in series) or $G + jB$ (conductance and susceptance in parallel) form, as in Figure 6 of Ref. 1. If they are in M $\angle$ ϕ (magnitude and phase) form, Figure 8 (ibid.) shows how they may be added graphically. However, Figure 24.1 shows that linear real and imaginary axes (as used in Ref. 1) do not provide a simple way of converting series resistance and reactance to the parallel form, or of converting it to $G + jB$ form. Further, really large values of R or X cannot be shown, as infinite values of these are off the page infinitely far to the east, or the north or south. There is a mathematical transformation of axes which solves both of these problems, and turns out to have other

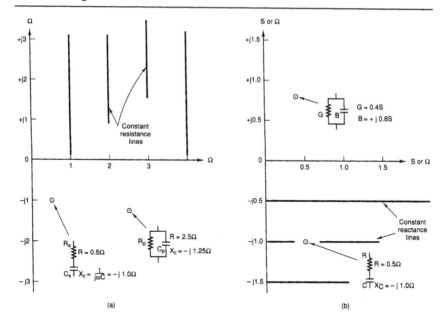

Figure 24.1 *(a) Series and parallel versions of the same lossy capacitive imped-ance plotted on real and imaginary axes. (Note: a particular radian frequency w is assumed.) (b) An impedance and the corresponding admittance plotted on the same real and imaginary axes. In neither (a) nor (b) is there a simple construction which will derive the one from the other. (Also shown in the diagrams are lines indicating loci of constant resistance, and of constant reactance.)*

extremely useful features as well. To bring the point representing infinite resistance onto the page, start by plotting values of resistance from zero to 1Ω on a base which becomes more compressed as it nears unity. Now for values greater than unity, plot the reciprocal of resistance to the same scale, so that infinity comes as far to the right of the point 1 as zero is to the left, see Figure 24.2 a). Next, gather up the tops and bottoms of the constant resistance lines (at infinity) and bend them round to the right, so that they become circles terminating at the (now unique) point representing infinity, at the right-hand end of the diagram, Figure 24.2(b)(i). In the process, the constant reactance lines have become arcs as shown in Figure 24.2(b)(ii), we have in fact a 'Smith Chart'. The constant resistance lines form an 'orthognal set' with the constant reactance lines: each line of the one sort crosses every line of the other sort at right angles. The constant resistance circles all have their centres on the horizontal diameter of the diagram, whilst the constant reactance (or susceptance) lines are all arcs of circles with their centres on a vertical line running from minus to plus infinity through the right-hand end of the diagram's horizontal diameter.

This chart can do double duty: it can represent zero resistance at the left of the horizontal diameter up to infinite resistance at the right, together with purely inductive reactance increasing from zero to infinity clockwise around its upper rim and capacitive reactance likewise anticlockwise around its lower. But equally, the horizontal diameter can represent zero conductance (open-circuit) at the left-hand end, up to infinite conductance (short-circuit) at the right; the upper arc now becomes capacitive susceptance from zero to infinite clockwise and the lower arc zero to infinite inductive susceptance anticlockwise. Furthermore, a very simple construction now takes one from a series value $(R+jX)$ of impedance to the corresponding parallel $(G+jB)$ admittance. Imagine any point on the chart (point A in Figure 24.2(c)), representing an impedance $(R+jX)\Omega$, then the point at the same distance from the centre and diametrically opposite gives the corresponding value of $(G+jB)$S. (S – for siemens – has replaced the mhos of my youth. Thus 2Ω is no longer half a mho but 0.5S, and 0S is an open circuit. Note that a capital S is used, as with F for farad, C for coulomb etc.; lower case s indicates the SI unit of time, the second.) In addition to providing a simple graphical conversion from series impedances to shunt admittances, the chart also provides a simple graphical means of finding the result of adding a series reactive component (inductive or capacitive, i.e. positive or negative) to an impedance, as shown in the lower half of Figure 24.2(d). Moving round the circle in this way is just the same as moving up or down a constant resistance line in Figure 24.1. (Adding more and more inductance would eventually take one right round through the top half of the chart, to the point of infinite reactance at the right-hand end of the horizontal diameter.) Similarly, adding a parallel susceptance to the starting admittance shown in the upper half of Figure 24.2(d) moves its effective value around the constant conductance circle $G = 0.5$.

There are two points it is essential to bear in mind when using the Smith Chart. The first is that one will either be using series components expressed in ohms or parallel components expressed in siemens, never parallel values in ohms such as those shown in Figure 24.1(a), or series values in siemens. Thus an impedance Z_1 $(=R_1+jX_1)$ in series with Z_2 gives a resultant $Z_{tot} = (R_1 + R_2) + j(X_1 + X_2)$, so $10+j200$ (an inductor with a Q of 20 at some particular frequency) in series with $90-j90$ (a capacitor so lossy that $\tan\delta = 1$!) gives $Z_{tot} = (100+j110)\Omega$, like a very lossy inductor or maybe a wirewound resistor. In like manner, an admittance Y_1 $(=G_1+jB_1)$ in parallel with Y_2 gives a resultant $Y_{tot} = (G_1 + G_2) + j(B_1 + B_2)$: e.g. a capacitive admittance $0.01+j0.1$ in parallel with an inductive admittance $0.01-j0.1$ gives $(0.02+j0.0)$S, an excellent match in a 50Ω system at one particular frequency (and a bit either side, as the Q is only 5). It is unnecessarily messy to work out the parallel combination of Z_1 and Z_2 above – much easier to convert them to Y_1 and Y_2 first. The second point is that the chart is drawn in terms of normalised impedances/admittances.

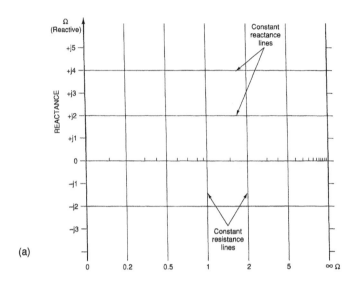

(a)

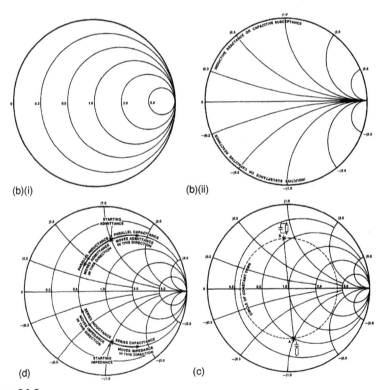

(b)(i)

(b)(ii)

(d)

(c)

Figure 24.2

This means that the impedance to which one may wish to match, say, the input of an amplifier, is assumed to be 1Ω. Having found the required normalised values of the components of the matching network, they are converted to those for, say, a 50Ω system by multiplying all inductive and capacitive reactances by 50 and dividing all the susceptances by 50; or by 75 in the case of a system with a characteristic impedance of 75Ω. Knowing the reactance or susceptance of each component, then given the desired operating frequency, its actual value is defined.

As an example of the Smith Chart in action, imagine it is desired to match the input of an IC which, at the desired operating frequency, looks like a normalised admittance of (0.71 + j1.72)S. This corresponds in a 50Ω system to a resistive component of 1/0.71 × 50 = 70.4Ω in parallel with a capacitive reactance of 29.1Ω – the capacitive term in the normalised admittance has a +j sign as it is a susceptance, the reciprocal of a capacitive reactance. The simplest course would be to add a shunt inductive susceptance of −j1.72, this value can be read off round the edge of the chart where the constant reactance (susceptance) lines are labelled. This would resonate out the capacitance, moving the point A in Figure 24.3 anticlockwise round the line of constant conductance to the point 0.71 on the horizontal axis. But a pure resistance of 70.4Ω, whilst an improvement, is not a perfect match to a 50Ω source: we need to *modify the effective resistive component* as well as remove any residual reactive component. This can be done in two stages, the first being the addition of inductance as before, but this time *series* inductance. To do this, it is necessary first to convert the starting parallel admittance point A to the series impedance form, point B, as described earlier. Series inductance can now be added to bring us to point C. The trick is to choose a point C which is diametrically opposite a point D on the constant conductance line G = 1 and at the same distance from the centre of the chart. (Point D represents exactly the same admittance as C, it is just that the latter expresses it in series impedance terms.) The required value of inductance is represented by the length of the arc BC, the values at B and C, read off from the edge of the chart, being −j0.5 and +j0.43 respectively. Thus a normalised series

Figure 24.2 *(a) The resistive axis modified to show all values from zero to infinity. (b) All points representing infinity (off the page to top and bottom, and on the constant resistance line at R = infinity) have been condensed into a unique infinity point at RHS of the diagram. (i) Constant resistance lines have become circles. (ii) Constant reactance lines have become arcs. (c) Showing how simply series impedance values are converted to parallel admittance values on the Smith Chart, compared with the calculations shown in the Box. (The significance of the dotted circle concentric with the centre of the chart is covered later in the text.) (d) Adding series inductance or shunt capacitance moves the resulting complex impedance or admittance clockwise around a constant resistance (conductance) line whilst series capacitance of shunt inductance moves it anticlockwise.*

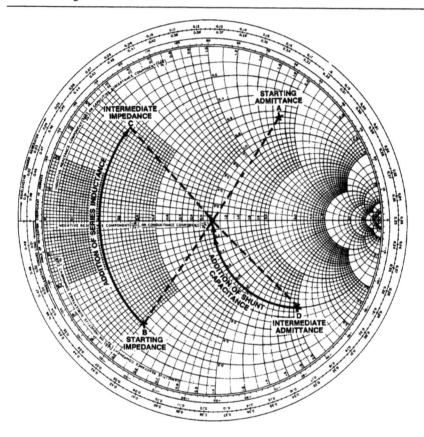

Figure 24.3 *Example showing the use of the Smith Chart to match a load to the source using lumped components.*

inductive reactance of 0.93Ω is required. Expressed as a parallel circuit, the input of the IC plus the series inductance looks like the admittance at point D, and adding the shunt capacitive susceptance indicated by the arc from D to the centre of the diagram, in this case $+j2.0$, completes the process of matching. The clever part is locating the point C and hence D. Lay a graduated straightedge (I use a transparent ruler marked in cm and mm) across the diagram, passing through the centre. Now swivel it round until the distance from the centre of the diagram to the intercept on the constant resistance line which passes through B, is equal to that on the constant resistance line passing through the centre of the diagram – note that as one intercept increases, the other decreases. If you repeat the exercise starting with a point $A' = (0.4 - j0.8)S$ (representing an inductive-ish input admittance looking like 125Ω in parallel with an inductive reactance of 62.5Ω), you will find that from point B' it is necessary to proceed anticlockwise adding series capacitance to arrive at a point C' which is diametrically opposite a point D' on the unity conductance circle and equidistant from the centre of the chart. Now, adding shunt capacitance as before (but rather less of it this time) brings the point to the centre of the diagram, the point of normalised impedance or admittance $1 + j0$. So where the first example required a matching network of series L and shunt C, this one needed two capacitors. Starting with other admittances could require series C and shunt L or even two inductors.

The foregoing explanation of the construction and use of the Smith Chart has been based around the assumption that the circuitry in question operates at a frequency where lumped components – capacitors and inductors – are appropriate, e.g. up to VHF and low UHF. However, at much higher frequencies, the values of lumped components may be inconveniently small, so it may be preferable to use lengths of transmission lines instead. Ref. 2 describes how a transmission line less than $\lambda/4$ in length at the frequency of operation, and with its far end open-circuit, looks like a capacitance, or an inductance if the end is short-circuited. The capacitive susceptance of the open-circuit line varies from zero when its length l is zero, up to infinity when $l = \lambda/4$ and the inductive reactance of the short-circuit line varies in just the same way. The voltage standing wave ratio on such a line of length $\lambda/4$ or more (the ratio of the maximum voltage on the line to the minimum – VSWR for short) is infinity, and similarly for any other purely reactive termination. On the other hand, on a line resistively terminated with the line's characteristic impedance Z_0, the VSWR is unity. If the shunt resistive component of the termination is not equal to Z_0, or if there is also a reactive component, then there will be some energy reflected from the end of the line, as a wavefront travelling back towards the source. Thus the voltage on the line will vary with distance from the termination: for example, if the voltage of the reflected wavefront is 10% of the incident, the voltage will vary between 1.10 where the incident and reflected voltage are in phase to 0.9 where they are in antiphase, giving a VSWR of 1.1/0.9 to unity or 1.222 : 1. The variation of reactance along a short-circuited line, moving away from the shorted end towards the source, can be plotted clockwise around the edge of the Smith Chart, from from zero (short) at the left-hand end of the horizontal diameter, through X_0 (an inductive reactance numerically equal in ohms to Z_0) at the top of the chart on to infinity (open) at the right-hand end (corresponding to a distance $\lambda/4$ along the line), and on again through a capacitive reactance equal to Z_0 to a capacitive short-circuit back at the starting point, at a distance $\lambda/2$ along the line. The outer edge of the chart is thus a circle of constant VSWR, namely infinity : 1. Smaller circles, concentric with the centre of the chart, represent lower VSWRs, right down to unity at the centre itself. The impedance seen looking into a line at an increasing distance from an arbitrary finite termination other than Z_o is given by following clockwise round a circle of constant VSWR passing through the point representing the termination. Armed with these results, plus the earlier ones concerning the addition of series and shunt compo-nents, the matching of a load to the source using lines is straightforward.

Figure 24.4 shows a Smith Chart with a load (consisting of resistance and capacitance in parallel) in normalised form of $(0.2 + j0.4)$S marked in, point A. Moving a distance of $(0.187 - 0.062)\lambda = 0.125\lambda$ towards the source brings us to the point B where the admittance consists of a conduc-tance 1.0 in parallel with a susceptance $+j2.0$. (Continuing around the chart – forwards or backwards – on a constant VSWR circle to point C tells one that without matching, the VSWR on the line would be $1/0.175 = 5.7 : 1$). Remembering that shunt admittances add directly, if we add a susceptance of $-j2.0$ across the line at a point 0.125λ from the load, it will

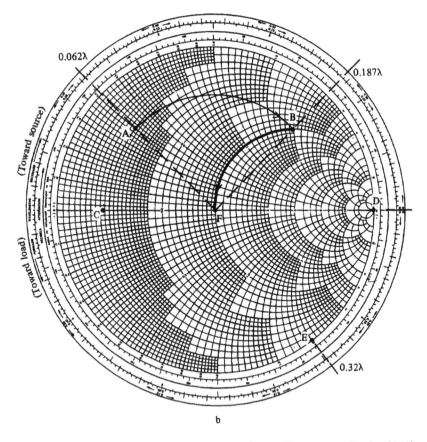

b

Figure 24.4   *Example showing the use of the Smith Chart to match a load to the source using lines. (Reproduced from Ref. 3 by courtesy of the publisher.)*

cancel out the susceptance of +j2.0 at B. In fact, the inductive shunt susceptance of −j2.0 parallel-resonates with the +j2.0 capacitive suscep-tance, so that viewed from the source, point B is moved around the constant conductance line to point F, representing a perfect match. The −j2.0 shunt susceptance can be a 'stub', a short-circuit length of trans-mission line. Point E represents −j2.0 susceptance and the required length of line, starting from the short circuit (infinite inductive susceptance) at D is $(0.32 − 0.25)\lambda = 0.07\lambda$. Thus connecting a short-circuit stub of length 0.07λ in parallel with the main line at a distance 0.125λ from the load completes the process of matching. This example of matching using lengths of transmission lines ignores the effects of any losses in the lines. This is permissible in practice as the lengths involved are so small. When working with coaxial lines, a short-circuit stub is usually preferred to an open-circuit, as it is more 'ideal', reflecting all of the incident power. An open-circuit can radiate a little of it, the more so the higher the frequency, resulting in a finite rather than a zero return loss. In microstrip and stripline, however, open-circuit stubs can readily be employed.

The Smith Chart illustrated in Figure 24.4 also shows various other parameters around the edge of the chart, in addition to the labels for the constant reactance/susceptance arcs, and the distance in wavelengths along the line. These additional items are the angle of the reflection coefficient, and attenuation. The latter is shown on an arbitrary scale of 10dB per half wavelength. If, for example, on an actual line, the attenu-ation is known to be 0.1dB/wavelength, then the attenuation read off from the chart for a given length of line should be divided by the figure $10/0.05 = 200$.

Box

The conversion of a series resistance plus reactance circuit into the equivalent parallel components and vice versa, is shown in the diagram. It summarises the results of some fairly straightforward algebra. To get from parallel impedances to parallel admittances (since the real parts of these – the conductances – add directly, as do the imaginary or suscep-tance parts), simply note that the in-phase and reactive parts G_p and B_p of the parallel admittance $Y_p = G_p + B_p$ are given by $G_p = 1/R_p$ and $B_p = 1/X_p$. Thus:

$$G_p = R_s/(R_s{}^2 + X_s{}^2) \quad \text{and} \quad B_p = X_s/(R_s{}^2 + X_s{}^2)$$

Similarly, turning the handle on the algebra:

$$R_s = G_p/(G_p{}^2 + B_p{}^2) \quad \text{and} \quad X_s = B_p/(G_p{}^2 + B_p{}^2)$$

The two sets of formulae have exactly the same form but with R and G changing places, and X and B doing likewise.

$Z_s = M_s \angle \phi_s$

$M_s = \sqrt{R_s^2 + X_s^2}$

$\phi_s = \tan^{-1} \dfrac{X_s}{R_s}$

$\mathcal{R} \qquad \cos\phi_s = \dfrac{R_s}{\sqrt{R_s^2 + X_s^2}} = \dfrac{R_s}{M_s}$

$\mathcal{I} \qquad \sin\phi_s = \dfrac{X_s}{\sqrt{R_s^2 + X_s^2}} = \dfrac{X_s}{M_s}$

$Z_p = M_p \angle \phi_p$

$M_p = X_p R_p / \sqrt{R_p^2 + X_p^2}$

$\phi_p = \tan^{-1} \dfrac{R_p}{X_p}$

$\cos\phi_p = \dfrac{X_p}{\sqrt{R_p^2 + X_p^2}} = \dfrac{M_p}{R_p}$

$\sin\phi_p = \dfrac{R_p}{\sqrt{R_p^2 + X_p^2}} = \dfrac{M_p}{X_p}$

For equivalence, $M_s = M_p$ and $\phi_s = \phi_p$

Serial to parallel,

$$R_p = \dfrac{R_s^2 + X_s^2}{R_s}, \qquad X_p = \dfrac{R_s^2 + X_s^2}{X_s}$$

Parallel to serial,

$$R_s = \dfrac{R_p X_p^2}{R_p^2 + X_p^2}, \qquad X_s = \dfrac{R_p^2 X_p}{R_p^2 + X_p^2}.$$

Figure A   *Showing the parallel resistance and reactance equivalent to a given series resistance – reactance combination, and vice versa. (Reproduced from Ref. 3 by courtesy of the publisher.)*

Acknowledgements

Several of the illustrations in this article are reproduced by courtesy of GEC Plessey Semiconductors Ltd, from the Application Note 'The care and feeding of High Speed Dividers', which appears in their Personal Communications IC Handbook Publication No. PS2123 June 1990. Note: this Application Note does not appear in the later version of the Handbook, HB2123-2, dated May 1992.

References

1. Bishop, O. 'Imaginary numbers for a real world', *Electronics World and Wireless World*, July 1993 pp. 604–612.
2. Hickman, I. Design Brief 'RF reflections', *Electronics World and Wireless World*, Oct. 1993 pp. 872–876.
3. Hickman, I. *Newnes Practical RF Handbook*, Butterworth-Heineman Ltd 1993 ISBN 0 7506 0871 4.

25 Magic numbers in electronics

The ancient Greeks knew about the golden ratio, pi and $(\sqrt{5}+1)/2$. But although Pythagoras and Euclid never had the opportunity to apply their work to filter design, those magic numbers crop up in second order filters! Several readers subsequently wrote to me, and a reply to them was published in the November 1994 issue of *Electronics World*, p. 950. It is reproduced below, following the article.

Ever since man started to count, numbers have fascinated him. Starting with the positive whole numbers (up to ten, perhaps, initially) primitive man at some point realised that there is no largest number, and eventually came to realise that there were other 'numbers' in between the whole numbers he was used to. For instance, whilst rulers of lengths 3, 4 and 5 cubits would let him build nice tidy right angles at the corners of a palace or house, the circumference of a barrel obstinately refused to equal a whole number times the diameter, although twenty-two sevenths seemed to be near enough for most practical purposes. Nowadays pi crops up in technical contexts, in electronics as elsewhere, for instance in $\mu_0 = 4\pi 10^{-7}$, the permeability of free space.

The sophisticated readers of this journal will be conversant not only with pi but also with e, the base of exponential or Napierian logarithms. Like pi, e is a truly magic number, popping up all over the place. It is also a number to be wary of – the exponential function has a dangerous tendency to explode. For instance, suppose that in 1066 near Hastings, one of William the Conqueror's soldiers wantonly did £1 worth of damage to the property of a local landowner, and that the landowner's descendants today obtained a court order for the payment of this sum with interest at a modest rate of, say, 2.5% per annum compound. Then the successors in title of William the Conqueror face a bill of eight thousand nine hundred and forty eight million, four hundred and thirty four thousand eight hundred and ninety eight pounds. If instead, 1.25% interest had been added six-monthly the figure would have been slightly higher. If interest had been added not six-monthly, monthly or even daily, but one millionth of the annual rate added every 31.5 seconds (one millionth of a year), the figure would have been slightly higher still, the effective annual rate becoming 2.531%. The total, instead of increasing in yearly steps, would have mounted up following a smooth curve described virtually exactly by an exponential function, in this case:

$$\text{Total} = \text{Principal} \times e^{0.025t},$$

where t is in years. You can imagine how rapidly the exponential function explodes if t is in seconds or even microseconds, especially if 'a' in e^{at} were unity or larger, rather than 0.025.

That an exponential cannot go on growing for ever is well known to engineers, but completely unknown either to politicians (with their talk of continuous sustainable growth), over-optimistic business men, or the poor unfortunates taken in by cleverly disguised chain-selling schemes. Exponential decay is much more well behaved, the voltage across a parallel resistor/capacitor combination dying away, like the world, with a whimper according to the equation

$$V = V_0 e^{\frac{-t}{CR}} \qquad (1)$$

This tells you what voltage is left across the capacitor t seconds after some arbitrary time of observation t_0 at which the voltage across the capacitor was V_0. Assuming that originally the capacitor was charged up to some enormous voltage, you can find out what the voltage was at any time *before* t_0 by letting time run backwards. Just substitute −t for t in equation (1), converting it into a positive exponent and a growing exponential.

In equation (1), the variable is time, but e appears in other equations where the variable is squared. I can't think off-hand of any equations where the variable is time squared, but other equations with e to the power (a variable squared) often occur. Naturally, to avoid explosions, the squared variable has an associated negative sign, just as time does in equation (1). Because there is no difference between x^2 and $(-x^2)$, a function defined by such an equation dies away as the variable increases in either a positive or a negative direction away from the norm or mean, as in the following equation, describing Gaussian noise – noise having a Gaussian or 'normal' distribution:

$$\text{Probability density of voltage } V = K_1 e^{-K_2 V2}$$

There is no maximum value to this function: in theory you could get a voltage spike of near infinite magnitude, but as the probability of this is near zero, you would have to wait for ever for the chance to observe it.

Incidentally, the same equation governs something as mundane as the tossing of a coin, at least if you do it often enough. Figure 25.1 shows the possible outcomes of tossing a coin ten times in a row. Interestingly, five heads and five tails is not the most likely outcome. Six of one and four of the other (not specifying which) is much more likely, though five of each is marginally more likely than six heads and four tails (or than four heads and six tails). If you make a histogram of the number of possible ways of getting 0, 1, 2 heads, the result closely resembles the distribution of Gaussian noise shown in Figure 25.2. As the number of tosses approaches infinity, the histogram converges ever more exactly on the normal curve.

We count in tens because we have five fingers on each hand. But 5 (or rather its square root) is the basis of another magic number which seems to be not at all well known, and which I will call K. This number is $K = (\sqrt{5} + 1)/2 = 1.618$. Clearly if a number is greater than one, its reciprocal is less than one, and vice versa. As it happens, K is the (only) number which differs from its reciprocal by exactly unity, so $K - 1 = 1/K = 0.618$, a number I shall call K'. Like e, K crops up all over the place. As Figure 25.3 shows, it describes the relative dimensions of a sheet of paper where the ratio of the short side to the long side is the same as the ratio of the long

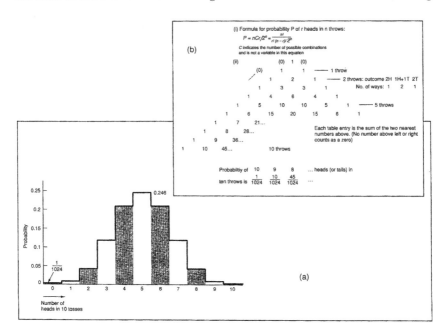

Figure 25.1 *(a) The odds of a tossed coin landing heads ten times in a row are 1 in 2^{10} or 1023:1 against. The histogram shows the probability for 0 to 10 heads as the fraction (number of ways of getting N heads)/(total number of possible outcomes) the denominator in this case being 1024. (b) The probability of r heads in n throws can be calculated either by formula (i) or by Pascal's triangle (ii).*

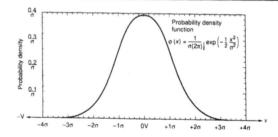

Figure 25.2 *The probability of the instantaneous value of random noise falling at any particular value is described by the normal curve, also known as the Gaussian distribution.*

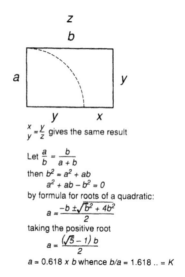

$\frac{x}{y} = \frac{y}{z}$ gives the same result

Let $\frac{a}{b} = \frac{b}{a + b}$

then $b^2 = a^2 + ab$

$a^2 + ab - b^2 = 0$

by formula for roots of a quadratic:

$$a = \frac{-b \pm \sqrt{b^2 + 4b^2}}{2}$$

taking the positive root

$$a = \frac{(\sqrt{5} - 1)\, b}{2}$$

$a = 0.618 \times b$ whence $b/a = 1.618 .. = K$

Figure 25.3 *Derivation of the magic number K.*

side to the sum of the long and short sides. I have a sneaking feeling this is called the 'golden ratio', but can't find it mentioned in any of my maths textbooks or encyclopaedias. It is said to be the most aesthetically pleasing ratio for a sheet of paper, being in fact only slightly squarer than the long and lanky-looking foolscap. A4 differs in the other direction, being squarer then $K:1$, in fact $\sqrt{2}:1$ so that on halving it to A5, the ratio is still the same.

What I hadn't realised until a few years ago is that K and K' crop up in electronics: in connection with filters in particular. The equation defining the response of a second-order low-pass filter is

$$\frac{V_0}{V_i} = \frac{1}{s^2 + Ds + 1} \text{ generally, or}$$

$$= \frac{1}{(j\omega)^2 + j\omega D + 1} \text{ in the steady state} \tag{3}$$

where ω equals $2\pi f$, and f is the frequency in hertz. For convenience, make ω' the 'normalised' frequency, i.e. the actual frequency divided by the filter's cut-off frequency. Thus at half the cut-off frequency, $\omega' = 0.5$ and so on, keeping the sums simple. D represents the damping term, which determines how high the peak at the upper end of the passband is, relative to the response at 0Hz, before the response falls away into the stop band. If $D = $ zero, corresponding to a Q of infinity since $Q = 1/D$, then the peak reaches infinite proportions. Figure 25.4(a) shows the response of a second-order low-pass filter for various values of Q up to infinity. With the response at $\omega' = 1$ being infinite, one might expect that it would still be very large even an octave above or below this frequency. In fact this is not the case, even at $\omega' = K$ or K', distinctly less than an octave away. As Figure 25.4(a) shows, the response at a frequency K is K' and at K' is K, as you may verify for yourself (or see from the Box) by substituting the appropriate values of ω in equation (3), with D equal to zero. In fact, you will find that the response at K' is +K and that at K is −K', indicating no phase shift in

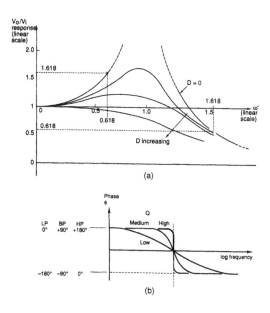

Figure 25.4 *(a) Amplitude response of a second-order low-pass filter. (b) Phase response of a second-order low-pass filter.*

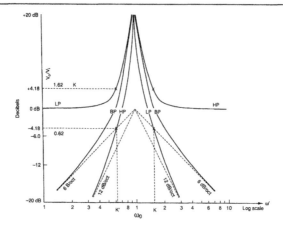

Figure 25.5 *Amplitude response (on a logarithmic scale) versus frequency (on a logarithmic scale) for low-, band- and high-pass filters, showing how the gain is K', 1 or K at frequency K' or K, according to type. Since K × K' = 1, these three points on the logarithmic frequency axis are equally spaced. K and K' on the amplitude scale correspond to ±4.179dB. Note: the bandpass curve always lies between the other two.*

the former case and 180° in the latter. For as Figure 25.4(b) shows, when Q = infinity there is no phase shift anywhere in the passband and the stopband phase shift is everywhere 180°. Amplitudes K and K' are sort of 'point asymptotes' or pegs in the ground. As the Q rises towards infinity the response at $\omega' = K'$ and K approaches these points, but can never exceed them.

The response of a second-order bandpass or high-pass filter is the same as in equation (3), but with s or s_2 respectively replacing 1 in the numerator of the right-hand side. You might expect K and K' to be involved here too, and you wouldn't be wrong. In the case of the infinite Q bandpass filter, K' and K are the frequencies where the response is j and −j respectively, i.e. the amplitude response is unity, the phase being 90° leading at K' and lagging at K. In the case of the infinite Q high-pass filter, the response at K' is −K' (−4.18dB and leading by 180°), and at K is K. This is shown in Figure 25.5, along with both the low-pass and bandpass results. The curves are not exactly to scale but are the right general shape. They are displayed on logarithmic axes, which permits the display of the 6 and 12dB/octave asymptotes as well.

Box

K and its stablemate K' show numerous relationships, which can be simply verified by algebra by substituting as appropriate. Note the following relationships:

$$1/K = K', \ 1/K' = K, \ K' + 1 = K, \ (K+1)/K = K, \ (1 - K')/K' = K'$$

For a second-order low-pass filter

$$\frac{V_o}{V_i} = \frac{1}{(j\omega')^2 + j\omega' \ D + 1} \text{ and if } D = 0, \text{ then at } \omega' = \frac{\sqrt{5}-1}{2} = K',$$

$$\frac{V_o}{V_i} = \frac{1}{\left(j\frac{\sqrt{5}-1}{2}\right)^2 + 1} = \frac{4}{-(5 + 1 - 2\sqrt{5}) + 4} = \frac{2}{\sqrt{5}-1} = \left(\frac{\sqrt{5}-1}{2}\right)^{-1}$$

$$= \frac{1}{K'} = K$$

For the second order bandpass case,

$$\frac{V_o}{V_i} = \frac{j\omega'}{(j\omega')^2 + j\omega' \ D + 1} \text{ and if } D = 0, \text{ then at } \omega' = \frac{\sqrt{5}+1}{2} = K',$$

$$\frac{V_o}{V_i} = \frac{\dfrac{j(\sqrt{5}+1)}{2}}{-\left(\dfrac{\sqrt{5}+1}{2}\right)^2 + 1} = \frac{\dfrac{j(\sqrt{5}+1)}{2}}{\dfrac{-(5 + 1 + 2\sqrt{5})}{4} + 1}$$

$$\frac{j(\sqrt{5}+1)}{-1 + \sqrt{5}} = -j = 1\angle - 90°$$

For the second-order highpass case,

$$\frac{V_o}{V_i} = \frac{(j\omega')^2}{(j\omega')^2 + j\omega' \ D + 1} \text{ and if } D = 0, \text{ then at } \omega' = \frac{\sqrt{5}-1}{2} = K',$$

$$\frac{V_o}{V_i} = \frac{-\left(\dfrac{\sqrt{5}-1}{2}\right)^2}{-\left(\dfrac{\sqrt{5}-1}{2}\right)^2 + 1} = \frac{\dfrac{6 - 2\sqrt{5}}{4}}{\dfrac{6 - 2\sqrt{5}}{4} - \dfrac{4}{4}} = \frac{\dfrac{3 - \sqrt{5}}{2}}{\dfrac{1 - \sqrt{5}}{2}} = \frac{1 + \dfrac{1 - \sqrt{5}}{2}}{\dfrac{1 - \sqrt{5}}{2}}$$

$$= \frac{1 - K'}{-K'} = \frac{1 - \dfrac{1}{K}}{-\dfrac{1}{K}} = \frac{K-1}{-1} = -K' = K'\angle + 180°$$

Magic numbers

I have received several letters concerning the golden mean, golden section or golden ratio mentioned in my article 'Magic numbers in electronics' (September, pp. 730–733). G C A of Dorset enclosed a photocopy of a brief entry in *Colliers Encyclopaedia* which gives the standard geometrical construction and states that the latter is dealt with in Euclid Book VI, *Proposition 30*. This led the Pythagoreans to a realisation of the existence and significance of 'incommensurables' or surds.

A C of Newcastle-upon-Tyne wrote to say that the Greek mathematician Eudoxus was the first man to investigate the golden mean, at first empirically and then more formally. His letter states that the Greek letter phi came to be associated with the golden mean after the artist Phidias, who used it extensively to proportion his sculptures. A C added that the ratio of successive numbers in the Fibonacci series approaches the golden ratio as the series progresses.

A H D of Hampshire wrote to say he was intrigued by the connection of the golden mean to the amplitude response of filters, but says it is not called K (or for that matter phi), but *tau*, being 'the Fibonacci number'. He kindly enclosed a photocopy of a four-page article which appeared in an issue of *Scientific American* of about a quarter of a century ago, concerning Fibonacci and his series, the first two terms of which are both 1, with each subsequent term being the sum of the previous two. It gives an explicit formula for the nth term, based upon K, phi, tau or call it what you will, and a long list of intriguing properties of the series.

Ian Hickman
Hampshire

26 Harmonising theory with practice

Textbook learning frequently leaves a gap between the theory you learn and the practical results viewed on an oscilloscope. This article seeks to clarify the relation between the time and frequency domain representations of common waveforms. (Note the error in Figure 26.3, top waveform. The second and fourth double spikes should go negative first, then positive.)

The competent electronic engineer carries a vast store of subject-related knowledge around with him, both theoretical and practical. Much of the former he will have learnt on a degree or similar course (possibly subsequently forgotten and then reacquired later as needed on the job) whilst much of the latter will have been picked up the hard way in the course of his work. Often, there is a bit of a gap between things that he knows on a theoretical basis, and the practical aspects of the same phenomenon, as viewed on, say, an oscilloscope or spectrum analyser. Years ago, curiosity about just such a dichotomy led me to look more closely at the relation between the time domain and frequency domain representations of some common waveforms, such as square- and triangular-waves, etc. The results are interesting, but are not spelt out simply in most textbooks.

Figure 26.1 shows a sinewave (top trace, angular frequency ωrad/s, say, or $\omega/2\pi$ cycles/s) and beneath it, its third harmonic at one third of the amplitude, its fifth at one fifth and its seventh at one seventh of the amplitude (thus relative to the fundamental, the amplitude of each harmonic is inversely proportional to its order). All these components start at the left-hand end of the plot at time $t = 0$, so that the angle ωt is also zero: all being sinewaves, this means that they all start from zero, positive going, as shown. The bottom trace shows their sum, and already a passable approximation to a squarewave is beginning to emerge. The 'flat' top of

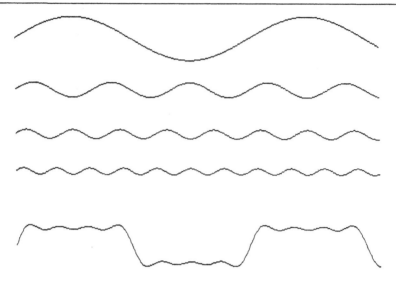

Figure 26.1   *(Top to bottom) Fundamental, with its third, fifth and seventh har-monics (at the appropriate amplitude and phasing) and their sum. As more har-monics are added, the sum approaches ever closer to an ideal squarewave.*

the squarewave has three dips and four bumps, indicating that only harmonics up to the seventh are present; quite a few more would be needed to make the top sensibly flat and, in particular, to make the rising and falling edges (at $\omega t = 0$ and $\omega t = \pi$ radians) vertical. Notice that the positive peak of the sinewave coincides with the *negative* peaks of the third and seventh harmonics (and the 11th, 15th, etc.) whilst it concides with the *positive* peak of the fifth harmonic (and the ninth, 13th, etc.). On the other hand, at $t = 0$ all are positive going, resulting in an infinitely steep rising edge, if you include odd harmonics all the way up to infinity, since the sum $1 + 1/3 + 1/5 + \ldots$ does not converge to a finite value but tends to infinity.

The top four waveforms shown are of course simply those predicted by Fourier analysis of a squarewave, and those predicted for a 'triwave' or triangular wave (at least, the first four again) are shown in Figure 26.2(a). Notice that this time, the positive peak of the sinewave coincides with the positive peaks of all the harmonics, resulting in the peak of their sum (bottom trace) becoming ever sharper as more harmonics are added. In addition to the altered phasing of the third, seventh, 11th (etc.) harmonics, this time the amplitudes of the third, fifth, seventh (etc.) harmonics are one ninth, one twenty-fifth and one forty-ninth (etc.) that of the fundamental, i.e. inversely proportional to the *square* of their order. So in Figure 26.2(a) the fundamental has been plotted at a larger amplitude than in Figure 26.1, so as to show the harmonics, which are now (especially the seventh and higher) of very low amplitude. And with the fundamental and even just

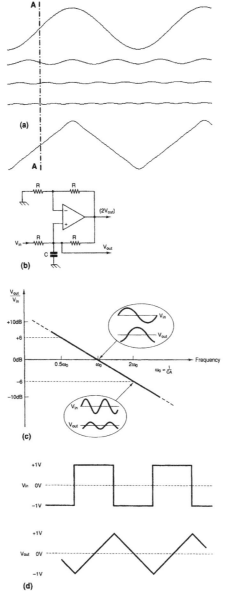

Figure 26.2 *(a) As Figure 26.1, except that the phasing and amplitude of the harmonics are those appropriate to generate a triwave. (b) An op-amp integrator circuit. This circuit behaves like a true integrator, its output voltage increasing when the input is positive, unlike the usual op-amp integrator which is in fact an inverting integrator. (c) Illustrating the integrator's frequency and phase response. (d) The effect of an integrator circuit upon a squarewave input whose half-period t = CR.*

a few harmonics, the triangular wave is now very convincing-looking – it just needs to be a bit sharper at its tips to be perfect.

So how does it come about that in the triwave the third and seventh (etc.) harmonics have become inverted in phase, whilst the fifth, ninth (etc.) haven't? To clarify this, let's switch for the moment to the time domain, and see how in practice the squarewave can be converted to a triwave. Figure 26.2(b) shows an op-amp integrator and Figure 26.2(c) illustrates its frequency and phase response. Its gain is unity at that frequency where the reactance of the capacitor in ohms equals the value of the input resistor R, falling at 6dB per octave above this frequency and rising at 6dB/octave below it (in both cases, for evermore – if the op-amp is perfect). In addition, an input sinewave (of any frequency) suffers a 90° phase lag in passing through the circuit: when the input sinewave is at its positive peak, the output is zero and increasing at its maximum rate. (The more usual sort of op-amp integrator is in fact an inverting integrator, so that the 90° phase lag looks at its output like a 90° *lead*. Figure 26.2(b) therefore shows a non-inverting integrator circuit, sometimes known as a de Boo integrator, which behaves exactly like an implementation of the mathematical oper-ation of integration.)

Figure 26.2(d) shows the response of the integrator to a squarewave input. All the while the input is positive, the output increases and likewise, when it is negative the output decreases. But to really understand the relation between the time and frequency domain representations, one must see how the effect of the integrator on the individual harmonics in Figure 26.1 results in the corresponding harmonic components in Figure 26.2(a). Note that compared to their phasing in Figure 26.1, the integrator has delayed all the frequency components in Figure 26.2(a) by 90° – at the left-hand side where t = 0 they are all at their negative peaks and don't pass through zero-going positive until 90° later. Now, for the fundamental, this corresponds to the time indicated at A–A, but at this point the third harmonic has moved through 270°, being at three times the frequency. Discounting the 90° phase shift suffered by the third harmonic itself, this means that, net, it has moved forward 180° relative to the phase relation with the fundamental at the input to the integrator. This is clearly shown in Figure 26.2(a), where at point A–A (corresponding to the left-hand side of Figure 26.1) when the fundamental is passing through zero positive going, the third harmonic is passing through zero *negative* going, resulting in its positive peak now coinciding with that of the fundamental. But the 90° delay of the fundamental at A–A corresponds to 450° at the fifth harmonic, or 360° discounting the 90° phase lag suffered by the fifth harmonic itself, so its phasing relative to the fundamental is unchanged. Thus now, the positive peaks of *all* the odd harmonics (not just those of the fifth, ninth etc.) coincide with that of the fundamnental, resulting in the sharp point seen in Figure 26.2(a).

Figure 26.3 illustrates the effect of repeated integration on a

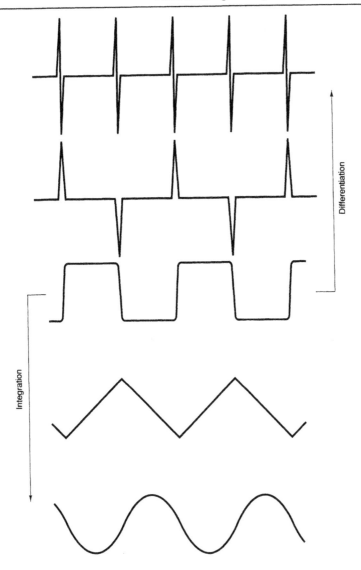

Figure 26.3 *The effect of repeated integration and differentiation on a (nearly) square wave.*

squarewave, and also of repeated differentiation, although in this case a finite rise time has been assumed, in order to avoid infinite amplitudes. Sticking with integration for the moment, Figure 26.4 shows the effect of a second integration of a squarewave, i.e. of integrating Figure 26.2(a)'s triangular wave. Again, the phase of the third and seventh harmonics (but not the fifth) has inverted, although at the screen resolution of Turbobasic

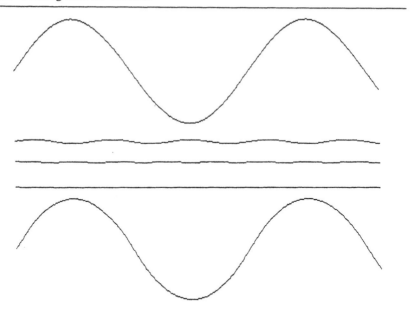

Figure 26.4   *Fundamental plus third, fifth and seventh harmonics of a triwave that has been integrated, and (bottom) the resultant.*

on a mono Hercules screen, the seventh is reduced to a straight line. Its amplitude is in fact $1/7^3$ times that of the fundamental, just -51dB or a more or less negligible 0.29%. At 1/27th of the fundamental, the third harmonic is responsible for 3.7% distortion, the total harmonic distortion being under 4%. Thus for uncritical applications, a twice integrated squarewave could stand in for a sinewave, although as can be seen by comparing the bottom trace in Figure 26.4 with the sinewave top trace, it is visibly just a little too rounded at the peak.

Figure 26.5 illustrates differentiation. As a differentiator has a frequency response which rises at 6dB/octave – just the very reverse of the −6dB/octave of the integrator illustrated in Figure 26.2(c) – the harmonics are emphasised so that they are now the same amplitude as the fundamental. Notice how in Figure 26.5, compared with the squarewave of Figure 26.1, all the frequency components have been advanced in phase by 90°. Thus like the triangular wave but unlike the squarewave, their positive peaks all line up with that of the fundamental. This is thus greatly accentuated into alternate positive and negative spikes, bottom trace, which as more and more harmonics of appropriate phase and amplitude were added would turn into infinitely high positive and negative spikes or 'delta functions'.

Returning to the squarewave, the relative amplitudes and phases of the fundamental and its harmonics are described exactly by an important

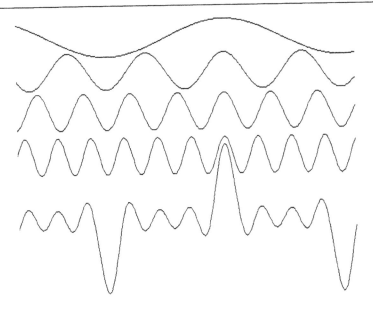

Figure 26.5 *When a squarewave is differentiated, the relative amplitudes and phases of the fundamental and the first three odd harmonics are as shown, adding to produce alternate positive- and negative-going spikes, bottom trace. Since, after differentiation, the amplitude of each odd harmonic of a squarewave is the same as that of the fundamental, many more harmonics would need to be shown to get very near the ideal waveform.*

mathematical function which, like π and exponential e, turns up all over the place. This is the expression

$$y = \frac{\sin(x)}{x}$$

When plotted, clearly it is going to look like a sinewave, but getting smaller and smaller on successive cycles, due to the x in the denominator. However, although sin(x) (and hence y) will generally be zero each time x is an integral multiple of π – i.e. at 180°, 360°, 540° etc. – 0° is a special case. Here, y = 0/0, a sum that your computer might spend a long time trying to work out, unless the software you are running is fitted with a trap and error message for 'divide by nought'! The standard way to evaluate a function at a point where its value is the quotient of two noughts is De l'Hospital's rule, but in this case it can be done by mental arithmetic, remembering that for x << 1 rad, sin(x) approximately equals x. The smaller x is, the more nearly exact is the approximation, so that as x tends to zero, y tends to 1. Plotted out, the function looks as shown in Figure 26.6(a), which for

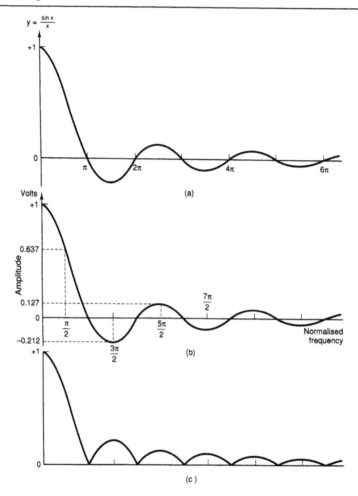

Figure 26.6   *(a) The 'sign x upon x' function, plotted for positive values of x only. (b) Showing how the value of y gives the amplitude and phase of the various frequency components of a squarewave when the frequency of the fundamental is represented by the value π/2. (c) The modulus of (a), i.e. showing only the numerical value of the function at each point, regardless of its sign.*

reasons of space has been plotted only for positive values of x: negative values would continue the curve to the left, mirror-image-wise about the vertical axis at t = 0.

Traditionally, x is used as the variable in this function, rather than, say, θ with good reason. The latter suggests an angle, whereas the variable may often be something different from this. In the case of squarewaves of all sorts, a useful representation results if one lets x represent not the instantaneous phase angle of the fundamental, but rather the various *frequencies* involved. The first zero of the function (i.e. when y = 0) occurs when x = 180° or π rad (Figure 26.6(a)). Now the positive half cycle of a squarewave is identical in shape to the negative half cycle, in the sense that if you flip the positive half down below the horizontal and then slide it along half a cycle, it fits exactly. This indicates that the waveform contains no even harmonic components, a point which you can verify for yourself by adding waveforms graphically as in the illustrations here, or by referring to the section on Fourier analysis in your college maths textbooks, which I hope that (like me) you have kept. So, what happens if we let the point x = π represent the frequency of the waveform's (missing) second harmonic component? The frequency of the fundamental would then be π/2, of the third harmonic 3π/2, of the fifth 5π/2 etc. The corresponding values of y are then [sin(π/2)]/π/2 = 2/π, one third of this value, one fifth and so on – precisely the relative amplitudes of the fundamental and harmonics of a squarewave.

Here, x represents the radian frequency, often alternatively called Φ, and x or Φ equals 2πf, where f is the frequency in hertz or cycles per second. So a radian frequency x = π/2 corresponds to (π/2)/2π, or 0.25Hz, but the curve can still represent any frequency squarewave you like, simply by the introduction of a suitable scaling factor. Likewise, a scaling factor can be used to adjust the amplitude y to represent the actual amplitude of any particular squarewave. Thus the curve can represent the constituent frequency components of any squarewave in both frequency and amplitude – and notice that its negative loops show the phases of the third and seventh harmonics, etc., to be opposite to that of the fundamental, fifth and ninth etc.: this is shown in Figure 26.6(b). The curve is sometimes seen drawn as in Figure 26.6(c), which represents only the amplitudes of the harmonic components, not their phases. This ties up with the picture you might see on a spectrum analyser, which shows only the relative amplitudes of the components of a complex waveform, but provides no information as to their relative phases.

The curve of Figure 26.6(a) actually also fits the spectrum of asymmetrical squarewaves and pulse trains of all sorts, and I hope to illustrate this – including its implications for DACs for digital audio – in another installment.

27 Analysing waveforms in the practical world

This article continues the investigation started in the previous one. Real live waveforms and advanced equipment are used to check how practical waveforms match up to their theoretical shapes – and to their spectral make-up as predicted by the (sin x)/x function.

Whilst last month's Figure 6(b) (reproduced here as Figure 27.1(b) correctly shows the relative amplitudes and phases of the harmonics of a squarewave relative to the fundamental component, it may seem odd at first sight that it gives the amplitude of the fundamental itself as 0.637, or 63.7%, or −4dB. An amplitude of 63.7% of what? – certainly not of the amplitude of the squarewave. If you sum $1 - 1/3 + 1/5 - 1/7 + 1/9 \ldots$ to infinity you get 0.782 as the amplitude of the squarewave whose fundamental component is unity. (Incidentally, you can work out this sum-to-infinity easily, if approximately, in a few moments on your calculator. Work out the sum up to and including the 11th harmonic, then add 1/13 to give the sum to the 13th harmonic. Add it to the sum to the 11th and divide by two. The same procedure using the 15th and 17th or even higher sums will give ever closer approximations.) Taking the reciprocal gives the fundamental component of a ±1V squarewave as ±1.278V. This brings us to the question of the reconstruction of sinewaves from their digital representation.

Figure 27.2(a) shows a digital audio system, where the ? block in the middle could indicate a compact disc, digital audio tape or digital compact cassette system, or even for the present purpose of explanation a straight-through wire connection. Figure 27.2(b) shows a sinewave input to the system and the same signal as it appears at the output of the DAC – assuming the ? block is just a piece of wire. Assuming the sampling frequency is one of the commonly employed ones, e.g. 44.1kHz, then the waveform shown (with exactly 48 steps per cycle) represents a frequency of

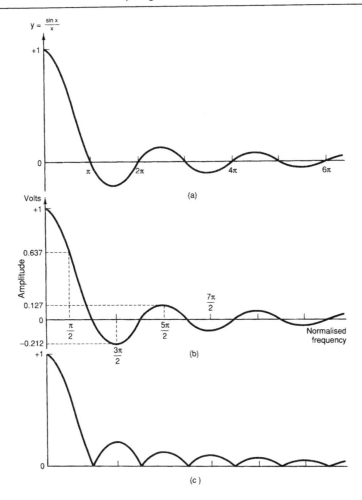

Figure 27.1 *(a) Sin x upon x. (b) Value of y gives the amplitude and phase of the various frequency components of a square wave. (c) Modulus of (a).*

just over 900Hz. Clearly, after low-pass filtering to knock off the ripple, it will be virtually an exact replica of the original, 100%, a gain of unity or 0dB down.

As is well known, a digital sampling system can represent frequencies (almost) right up to the 'Nyquist rate', or half the sampling frequency. Figure 27.3(a) shows the DAC output when the input to the ADC is just below the Nyquist rate, and at first sight it does not look very much like the input signal. But this is because it contains the image frequency component, which is just *above* the Nyquist rate. Figure 27.3(b) shows the first part of the waveform of Figure 27.3(a) to an expanded time scale, together with

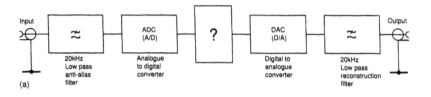

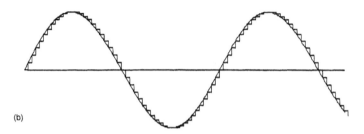

Figure 27.2 *(a) Basic digital audio signal chain. The ? block could represent a compact disc, digital audio tape or digital compact cassette storage. (b) A sinewave of around 1kHz, as it might appear at the output of the DAC, before low-pass filtering, shown with the original sinewave.*

the original sinewave. When the waveform of Figure 27.3(a) is passed through a low-pass filter with a steep cut-off at the Nyquist frequency, the image component is suppressed (along with the sharp edges, which represent harmonics of the wanted and image frequencies) leaving just the original sinewave shown in Figure 27.3(b). Assuming that the sinewave amplitude is ±1V pk-pk, you can see in Figure 27.3(b) that the squarewave out of the DAC varies in amplitude between zero and ±1V pk-pk. At its maximum, the fundamental component of the squarewave will be, as noted earlier, ±1.278V pk-pk. As the wanted frequency and its image are equal in amplitude, after the low-pass filter has suppressed the image and all the harmonics, the amplitude of the wanted signal will be 1.278/2 or ±0.637V pk-pk. Thus whilst the fundamental component of the 48-step representation of a sinewave shown in Figure 27.7(b) differs in amplitude from the original virtually not at all, a signal at (or at least very close to) the Nyquist rate will be, after filtering, only 63.7% of the true amplitude, or 3.9dB down, as predicted by the sin x upon x curve. Similarly, a frequency at a quarter of the sampling rate (or half the Nyquist frequency) will be 0.9dB down, and the expression sin x upon x predicts the loss at any other frequency you care to mention. This is why the better digital audio reconstruction chips include a 'sin x upon x compensation filter', to correct the mild high frequency roll-off which the signal would otherwise suffer, for

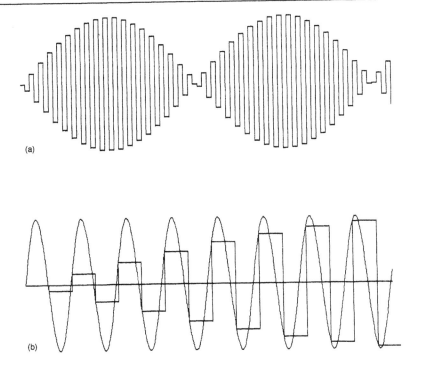

(a)

(b)

Figure 27.3 *(a) Waveform of the output of the DAC for a frequency just below the Nyquist frequency – not at first sight very much like the original sinewave. (b) The left-hand end of waveform (a) expanded timewise and showing the original sinewave as well.*

the benefit of all those golden ears which would otherwise be disappointed by the reduced level at 20kHz.

Asymmetrical fit

The (sin x)/x curve of Figure 27.1(a) actually also fits the spectrum of asymmetrical squarewaves and pulse trains of all sorts. In fact, an equal mark/space ratio pulse train is simply a squarewave with a DC component, as Figure 27.4(a) shows. If now the pulse-width of such a train is kept the same, but the space between pulses increased slightly, the frequency of the fundamental will be slightly lower than shown in Figure 27.1(b), so that its second harmonic will now fall just to the left of the first zero of y. This is shown in Figure 27.4(b): the waveform now has some even harmonic components. If the space between pulses were further widened to twice the pulse width, the frequency would fall to two thirds of that shown in Figure

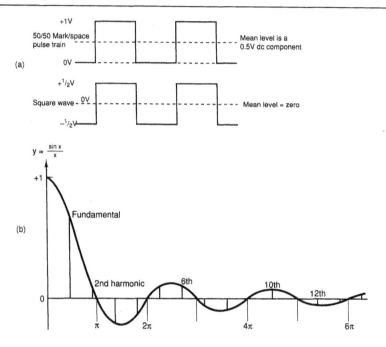

Figure 27.4   *(a) The only difference between a squarewave and a unipolar pulse train with unity mark/space ratio is that the latter has a DC component, the former does not. Apart from this, the spectrum of each is identical. (b) If the width of the pulses in (a) is kept the same but the distance between them is increased slightly, then the frequency is slightly lower, so the second harmonic will no longer fall on a zero of the sin x upon x function.*

27.1(b). The spectrum would then have both odd and even harmonic components, but the third, sixth, ninth harmonics, etc., would all be zero. If (still with the same pulse-width) the mark/space ratio were further reduced to 25/75%, the fundamental frequency would be half that shown in Figure 27.4(a) and it would be the fourth, eighth, 12th harmonics, etc., that would be zero.

Testing theory

At least, that's the theory, and the practice must tie up – mustn't it? Just to make sure, I built up the circuit shown in Figure 27.5(a), using fast 74ACT devices clocked at a leisurely 4MHz, and giving out a 250ns pulse every microsecond as shown in Figure 27.5(b). (The fourth section of the NAND gate was used to provide a delay to match that through the second flip-flop. Before adding it, a narrow sneak pulse or glitch occurred at the output

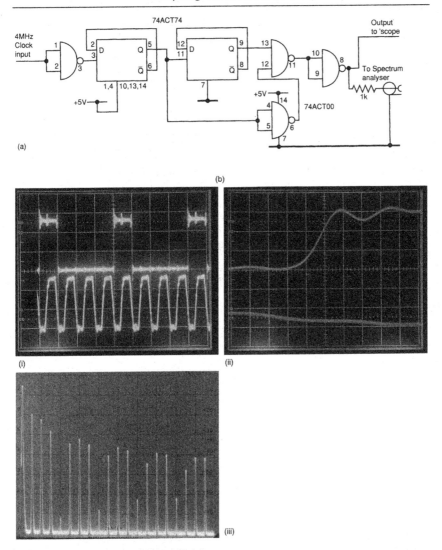

Figure 27.5 *(a) Circuit used to produce a pulse train with an accurate 25/75% mark/space ratio. (b) Waveforms associated with (a): (i) 4MHz CMOS input clock (lower trace) and 1MHz 25% duty cycle pulses (upper trace) both at 2V/div. 250ns/div.; (ii) leading edge of (i) at 1ns/div., showing the pulse rise time to be less than 2ns. (c) Spectrum of the pulse train produced by the circuit in (a), display centre frequency 10MHz, 2MHz/div. horizontal, ref. level 10dBm, 10dB/div. vertical, IF bandwidth 30kHz, video filter off.*

when the first flip-flop output went positive, triggering the second flip-flop's output to change from 1 to 0. This glitch was due to the propagation delay through the second flip-flop before its output fell to the 0 level, resulting in a couple of nanoseconds during which the NAND gate inputs at pins 12 and 13 were both positive. To get the ringing down to the level shown, considerable care in connecting the probe was necessary, discarding its usual earth lead in favour of a short thick wire strap to circuit ground. Similar precautions were taken in connecting the signal to the spectrum analyser.) The spectrum of the waveform is shown in Figure 27.5(c). The null at the fourth harmonic would look fairly convincing even on a linear scale, but note the log scale here – so it is actually well over 50dB down on the fundamental, or less than 0.3%. Feeding the values of frequency of the fundamental, second and third harmonics, namely $\pi/4$, $\pi/2$ and $3\pi/4$ respectively, into the formula $y = (\sin x)/x$ gives values of amplitude for these components of 0.90, 0.64 and 0.30. From these figures, the second harmonic calculates out as 3dB down and the third as 9dB down on the fundamental. This ties up pretty closely with the spectrum in Figure 27.5(c), which also shows the higher harmonics falling into the pattern shown in Figure 27.1(c) (allowing for the vertical scale being logarithmic, not linear). Thus the sin x upon x curve is an envelope or locus which passes through the values of amplitude of the component frequencies of any pulse waveform.

Reduced pulse width

The foregoing examples illustrate that if the width of the pulses or 'marks' is held constant but the width of the intervening spaces is increased, lowering the frequency, the positions of the zeros of the sin x upon x function are unchanged but the spectral lines become more numerous and closer together. Thus the *frequency* of the waveform determines the *spacing* of the spectral lines within the first loop of the sin x upon x function (and elsewhere), the lines being of course equally spaced everywhere. But what happens if one fixes the PRF (pulse repetition frequency) whilst reducing the pulse mark/space ratio by increasing the space at the expense of the mark? The PRF being fixed, the position of the lines on the x or ω axis will remain the same, but as the second (and other even harmonics) must start to appear, the first zero crossing of the sin x upon x function must move to the right – further and further to the right as the mark/space ratio becomes smaller and smaller. And as the pulse narrows down to infinitesimal width (which means its amplitude must be greatly increased if it is still to convey a finite amount of signal power), the first zero crossing moves out towards infinite frequency. Thus the spectrum of such a 'delta function' consists of a component at the fundamental frequency plus all its harmonincs, all in

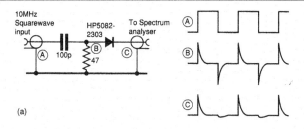

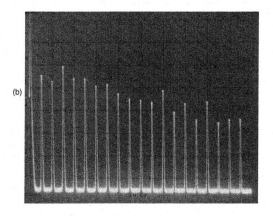

Figure 27.6 *(a) Producing some narrow pulses at a PRF of 10MHz. (b) The spectrum of (a) approximates a sea of equal amplitude 10MHz-spaced spectral lines stretching out towards infinite frequency. Centre frequency 100MHz, 20MHz/ div. horizontal, 10dB/div. vertical, IF bandwidth 100kHz, video filter max.*

phase and all of equal amplitude. This can be (approximately) demonstrated by differentiating a squarewave and then picking out just the positive-going edges with a fast diode as illustrated in Figure 27.6(a). The 10MHz squarewave was obtained from the same generator as used for the 4MHz clock in Figure 27.5, but this time using its much faster SN74128 50Ω line driver output. The time constant CR of the differentiator shown is 4.7ns or nearly 5% of the squarewave's period, so the resultant pulses won't be exceedingly narrow – just 4.7ns wide at 37% of full amplitude given a near-ideal squarewave drive – and of course they will be a different shape from that considered up till now. Nevertheless, the resultant spectrum, Figure 27.6(b), is clearly going in the right direction. The reason alternate harmonics (the even ones) are slightly smaller than the odd harmonics is incomplete suppression of the negative-going spikes by the diode. If the diode were shorted, feeding both positive and negative spikes to the spectrum analyser, the even harmonics would disappear completely; Figure 26.5 shows that spikes of both polarities involve no even harmonics.

A train of narrow pulses, i.e. a comb of near-equal amplitude sinewaves like this, has numerous uses, one example being as the LO (local oscillator) drive in a superhet receiver. The LO is arranged to run at a very much lower frequency than the band of interest, feeding narrow spikes to the mixer. Any signal within the bandwidth defined by the receiver's front-end bandpass filter is likely to be near enough to one of the comb of 'local oscillator frequencies' to be received. This permits a sensitive narrow-band (and hence low-noise) receiver to cover the whole band at once without the delay taken with a single swept LO approach – handy if you are sweeping a room for radio bugs.

Sinewave continuum

Now let's consider the case where not only does the pulse width tend to zero (pushing the frequency of the first zero of the sin x upon x function out towards infinity), but its frequency tends to zero also. The lower the frequency of the narrow 'LO' pulses, the closer together are the spectral lines. Thus ultimately the spectrum becomes an infinite number of sinewave components all of the same amplitude and so closely packed together as to represent a continuum.

In Fourier analysis, the sinewave components are assumed to exist from time equals minus infinity to plus infinity, so that the waveforms shown in Figures 26.1, 26.2(a), etc., extend off the page to left and right indefinitely. This is fine as long as the squarewave (or whatever) that they represent also exists permanently, but it can lead to some interesting but idle speculation when the PRF has fallen to zero, so that they represent an isolated pulse occurring at time t = 0. The miriad of frequency components of negligible amplitude can be pictured as all buzzing away, so phased that their sum is zero at every instant, except when the magic moment t = 0 arrives. At that point they are all uniquely in phase, resulting in just the one isolated infinitely narrow never-to-be-repeated spike.

No pulse paradox

Suppose you have set up a programmable pulse generator to give a single narrow delayed pulse so that, say, t = 0 is set for 3 o'clock next Tuesday afternoon. What would happen to all those components if, just before 3 pm there was a power cut so that unexpectedly the pulse did not after all occur? There is really no paradox here, for the existence of harmonic components of a single isolated pulse *prior to the pulse* is only a mathematical convenience – such precursors are 'non-causal'. If you could suddenly 'turn on' all those components, each at the same amplitude and the correct phase (all at zero, positive going), they would add up to give that single isolated pulse. They would also never all come in phase again simultaneously to produce another pulse. So, you might as well then turn them all off again – as 'postcursors' they are also ineffective or non-causal.

Key experiment

But that doesn't mean their combined effect does not linger on, as you can prove with a simple experiment which you can carry out yourself. Many homes contain an audio spectrum analyser of the filter-bank variety, with considerably better frequency resolution than the usual third-octave filters. If you hold down the loud pedal with a couple of bricks and take off the front (or raise the lid, if you run to a grand piano) the experiment can begin. Wait till all is quiet (let $t = 0$ be the middle of the night if necessary) and clap loudly, once. You will hear all of the piano strings sounding, each responding to its particular frequency component of the near-delta-function. The sound soon dies away, since the strings exhibit a finite Q; if they weren't losing energy in the form of sound waves you wouldn't be able to hear them. But if the Q of the individual filter-bank channels were near infinite, the post cursors would indeed continue (almost) for ever.

28 Sinewaves step by step

A reader wrote in to the Letters column of *Electronics World and Wireless World* with a query about the transient occurring at the initiation of a sinewave by a step function. This article was an attempt to clarify what happens, without resorting to the mathematics. The subsequent correspondence shows that confused thinking, neglecting factors like rise time, the series loss component in a capacitor or the self-capacitance of an inductor, can get one into an awful tangle. The parasitic parameters of real life components are a great inconvenience, but you ignore them at your peril!

Following one of my earlier articles, a reader wrote in ('Sine Post', Letters *Electronics World and Wireless World*, Feb. 1995, p. 149) with a query about the initiation of sinusoidal waveforms in a simple series L, C, R circuit, by the application of a step function of voltage to the circuit. Why should the voltage waveforms developed across the (lossy) inductor and the (ideal) capacitor always be sinusoidal? He goes on, in a sense, to answer his own question by acknowledging that sinewaves drop out in the solution of a second-order differential equation describing the system, but seeks a more intuitive insight based on the physical properties of inductors (terminal voltage proportional to the rate of change of flux linkage) and capacitors (terminal current proportional to the rate of change of stored charge).

Such an intuitive explanation is indeed possible, as will appear in what follows, though some readers may point out that it simply amounts to solving the differential equations by stealth. And to make the explanation simpler, the series circuit shown in Figure 28.1 assumes that r (the resistance of the inductor) is negligible – its Q infinite, an assumption which is also applied to the capacitor. Furthermore, the transient has been applied not as a voltage step function in series with the circuit, but as a unit current impulse or delta function in parallel with it, again in the interests of keeping everything as simple as possible for the purposes of explanation. After all, both a step function and a delta function are bony, angular, non-repetitive

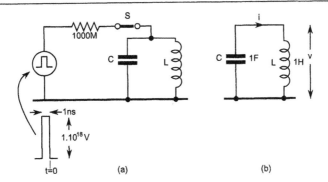

Figure 28.1 *(a) LC circuit with delta function excitation. (b) At time t = 0 and after, the situation is as shown.*

shapes, and about as far as one can get from a repetitive curvaceous sinewave.

The generator G in Figure 28.1(a), with its 1000MΩ internal resistance, provides an almost perfect delta function, dumping a current of 1×10^9A into whatever circuit it is connected to, for a period of 1ns. Thus the delivered charge is one coulomb, assuming the voltage in the target circuit remains negligible compared to 1×10^{18}V – not a difficult condition to meet. (The circuit values may seem a little impractical, but for the purposes of explanation, the theorist has the advantage over the practical man that he can choose whatever values he likes. Furthermore, his circuits always work, serenely untroubled by parasitics or other unforeseen contingencies, but – unfortunately – only on paper.)

As our interest centres on the response of the circuit to the transient stimulation – rather than on the transient itself, let t = 0 be the moment immediately following the delta function, when the generator voltage has just returned to zero. During the preceding one nanosecond it will have dumped a charge of 1 coulomb (1C) into the capacitor C, whose terminal voltage will have risen linearly over that nanosecond from zero to 1V. Left connected, the generator would then draw a current, albeit tiny, from the capacitor, so the switch S has been included in the circuit – we can open it at our leisure just after t = 0. The other point to clarify concerns the inductor L: at t = 0, the voltage across it is 1V, so the rate of increase of current through it is 1 amp per second (1A/s). But at t = 0, the current itself can be assumed with negligible error to be zero, despite the fact that there was some voltage across it during the one nanosecond period preceding t = 0, since the impressed volt-second product was negligible, see Box. All these seemingly nit-picking preliminaries are necessary not only to satisfy the rigour of the mathematical purists, but also for the more important practical reason of ensuring that we get the right answers.

So, at t = 0, there is zero current in the inductor and the instantaneous voltage v across the capacitor is 1V. The voltage across the inductor is proportional to the rate of change of flux linkage, and here (unlike the case of a generator) there is no externally applied flux. So the flux is simply proportional, at every instant, to the current through the inductor. Hence the rate of increase of current i in Figure 28.1(b) at t = 0 is 1A/s. As i increases, it will draw charge from the capacitor, resulting in a reduction of the voltage across the capacitor, according to the equations in the Box. If the voltage across the capacitor is regarded as the cause and the current through the inductor as the effect, then here, the effect is seen in its turn directly to affect the cause. This makes it difficult to see what is going on, which is why methods for solving differential equations were invented in the first place. However, we can see (approximately) what is going on by imagining the voltage across the capacitor held constant for a short period Δt while we see what happens to the current in the inductor, and then making a correction to the capacitor voltage to allow for the charge drawn off during that instant.

The Turbobasic program shown in Table 28.1 does just this, and should be largely self-explanatory. When run, it calculates out and plots the circuit voltage and current as shown in Figure 28.2. It shows the voltage and current indeed to be sinusoidal, and given that the 199 steps of 50ms shown correspond to just over 1.5 cycles, this ties up with the 0.159Hz predicted as the resonant frequency by the formula $f = 1/\cdot 2\pi\sqrt{(LC)}$. But the gradually increasing amplitude indicates a circuit with a Q of greater than infinity:

Table 28.1 *Turbobasic program to plot LC response to a delta function (unit impulse). It should run under Turbobasic on almost any pc XT/AT.*

```
100 REM ww_sine.bas
110 SCREEN 2: REM selects turbobas screen
120 WINDOW (0,399)-(399,0) REM redefines window to read 1 to r, bottom to top
121 h=2: REM sets horizontal step size
122 u=300: REM sets vertical position for voltage display
123 w=100: REM sets vertical position for current display
124 s=50: REM sets display vertical scale
126 i1=0: v1=1: REM initialises current i Amps and voltage v Volts
127 t=0.05: REM sets time step to 50ms
130 FOR j=0 TO 199
140 i2=i1+v1*t: REM calculates new value of current
160 v2=v1-(i1+i2)/2*t: REM adjusts capacitor v for average i drawn off
180 LINE (h*j, u+s*v1)-(h*(j+1), u+s*v2): REM plots lines joining present vals
185 LINE (h*j, w+s*i1)-(h*(j+1), w+s*i2): REM of v and i to previous ones
190 i1=i2: REM redefines present new value of current as next old value
191 v1=v2: REM as above, for voltage
200 NEXT
210 REM draw scale lines
220 LINE (0, 0)-(0, 399)
250 LINE (0, u)-(399, u): LINE (0, w)-(399, w)
280 LINE (0, u+50)-(399, u+50): LINE (0, w+50)-(399, w+50)
300 LINE (0, u-50)-(399, u-50): LINE (0, w-50)-(399, w-50)
```

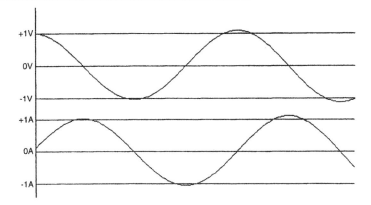

Figure 28.2 *Behaviour of the circuit of Figure 28.1(b), according to the program in Table 28.1.*

there must be something wrong somewhere! Line 140 calculates the increase in the currrent over a short period Δt (assuming the capacitor voltage remains constant the whiles), and line 160 then adjusts the capacitor voltage assuming the charge drawn off during Δt is equal to the average current during that period times Δt. Using the average current like this would be all right if, to be fair, the average voltage $(v_1 + v_2)/2$ had been used in line 140, instead of just v_1. But until line 160 has been calculated, the program can't know the value of v_2. The way round the difficulty is to use pairs of periods Δt, during the first of which v_1 is held constant giving a new value i_2 for i, and then during the second this new value of i is held constant, giving a new value v_2 for v. The only modification required to the program is that line 160 becomes $v_2 = v_1 - i_2t$. With this correction, the peak values of voltage and current remain constant, as would be expected for a tuned circuit with no damping, even if the time step is made much longer than the 50ms defined in line 127. The amplitude remains constant even if Δt is made as large as, say, 800ms, a significant fraction of the 6.28s period of the sinewave across the tuned circuit. This is shown in Figure 28.3, where you can see that there are $2\pi/0.8$ or nearly eight segments per cycle; that the theory of finite differences really works never ceases to amaze.

Using the present value of voltage to predict the new value of current, and then using the new value of current to predict the new value of voltage is a standard way of generating quadrature sinewaves such as those shown in the figures. It can be done to any desired degree of accuracy in DSP,[1] and it can equally well be done analogwise in hardware. In this case, the result is likely to be a stepwise approximation to a sinewave, as in an article which appeared in this journal some years ago,[2] rather than the linear

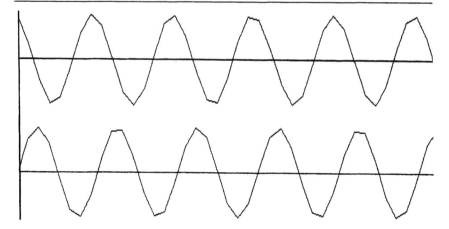

Figure 28.3   *Approximate behaviour of the circuit of Figure 28.1(b) predicted with the corrected version of the Table 28.1 program, but with a coarse time increment Δt, greater than a tenth of the period.*

interpolation representation shown in Figures 28.2 and 28.3. Figure 28.4, reproduced from Ref. 2, shows the sinewave generated at 500Hz, and Lissajous figures showing the very close approximation to quadrature achieved between the sine and cosine outputs both at 5Hz and 5kHz. The only essential difference between the circuit used in Ref. 2 and the program of Table 28.1 is that instead of using a transient as the excitation, the loop gain of the circuit was made slightly greater than unity so that the amplitude built up to the point where some diode gain-reducing networks came into play.

Returning to our starting point, the program in Table 28.1 is easily modified to include a finite loss resistance r in series with the inductor. Following t = 0, as the current increases, the volt drop across the resistor must be deducted from the capacitor voltage to find the voltage impressed across the inductor. The reduced volt-second product impressed on the inductor means the peak current will be reduced, and this will apply twice each cycle as the current flows first one way and then the other, the waveform dying away exponentially. And instead of the delta function generator of Figure 28.1(a), one could connect a 1V step function generator between the lower plate of the capacitor and ground in Figure 28.1(b). The waveforms would be unchanged, but after the waveform had died away, there would remain an amount of energy $(C.V^2)/2 = 0.5J$ (in this case) stored in the capacitor.

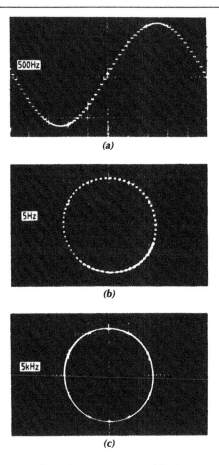

(a)

(b)

(c)

Figure 28.4 *(Reproduced from Figures 3 and 4 of Ref. 2).*

To the Editor

Dear Sir

On PD's own analysis, there cannot be an instantaneous voltage step across an ideal capacitor, unless, that is, an infinite current flows – which was precisely the case with the delta function in Ref. 1. If he really seems to see an 'instantaneous' voltage step across the capacitor in his circuit, there are a limited number of possible explanations:-

(i) The capacitor possesses significant series loss (possible, but not likely),
(ii) The capacitor possesses significant inductance (as PD himself suggests; was it a capacitor rated for pulse operation?),
(iii) The rise in voltage was not really instantaneous, or

(iv) There is a measurement error (it was not clear how he was measuring the voltage across the capacitor, which – from the diagrams supplied with his letter – had neither end grounded).

The analysis of the operation of his circuit is complicated by the fact that his 'step function' recurred, with alternating polarity, every 5ms; i.e. it was actually a 100Hz squarewave, whilst the resonant frequency of the tuned circuit to which it was applied was observed to be only some thirty times higher (approximately), with an unspecified Q. Furthermore, far from being instantaneous, the rise time of the 'step function' from zero to +4V was 50ms, around one sixth of the period of the tuned circuit's natural frequency.

A solution of the circuit's response to the given stimulation is straightforward, but could not be undertaken without exact values for the complex impedances of the components used – for example the iron-cored inductor doubtless had significant iron and copper loss in addition to its self-capacitance. (The values of L and C – 0.0016μF and 5H – which PD gives do not tally even approximately with his observed natural frequency of around 3kHz.) The observed voltage step across the capacitor is probably due to the division of the applied step between the said 0.0016μF capacitor and the self-capacitance of the inductor.

As the stimulus is a simple recurrent waveform, the circuit could be analysed in either the time or frequency domain, though of course both analyses would give the same result. However, one important point is perhaps more easily made clear by consideration in the frequency domain. The squarewave drive signal will have significant harmonics up to the resonant frequency of the tuned circuit. If the tuned circuit has a high Q and resonates exactly at one of these harmonics, there will be no phase changes in the damped oscillatory response, only amplitude changes. However, slight mistuning either side of the harmonic can result in dramatic changes in the response, as was illustrated with actual measurements in Ref. 3. If on the other hand, the circuit Q is so low that the response to one edge of the squarewave dies away completely before the arrival of the next, then analysis of the effect of an isolated quasi-step function with a finite rise-time would give the complete solution. Either way, there is no need to invoke unknown effects of the Ether to explain the observed results.

1. Hickman, I. 'Sinewaves – step by step', *Electronics World and Wireless World*, Mar. 1995 p. 215.
2. Hickman, I. 'Integrated creativity', *Electronics World and Wireless World*, Jan. 1992pp. 40–42.

Yours faithfully
Ian Hickman

References

1. Darwood, N. 'Accurate sinewave oscillator', *Wireless World*, June 1981.
2. Follett, D. H. 'Accurate sinewaves', *Wireless World*, Nov. 1981.
3. Hickman, I. 'Integrated Creativity', *Wireless World*, Jan. 1992, pp. 40–42, reproduced in 'Analog Circuits Cookbook', Ian Hickman, Newnes, 1995, ISBN 07506 2002 1.

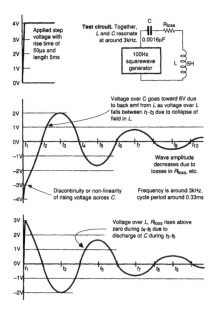

The circuit and observer results described by P.D. in a letter to the editor

29 Is matching easy?

On the face of it, matching the output of one circuit to the input of another is straightforward. The maximum power theorem indicates that the load can depart widely from 50Ω, yet the power delivered to it from a 50Ω source varies little. But when both source and load impedance depart from the ideal, problems really start.

Matching a load to a source (or vice versa) will ensure that the maximum possible power will be transferred from the one to the other, as stated by the well-known Maximum Power Theorem. In electronic signal processing, the matched condition is often therefore the preferred one, but not necessarily in other applications. For instance, the internal resistance (the 'source resistance') of a new 1.5V cell is around the 1Ω level, whereas the resistance of a 1.5V 300mA lens-end flashlight bulb is 5Ω (when lit). This ensures that five sixths of the energy supplied by the cell finishes up where it is wanted, producing light. In the matched case, a 1Ω bulb might produce more light, but 50% of the expensive energy you paid for when you bought the cell would be wasted simply warming it up. If you want the extra light, it is better sense to use more cells in series and a higher voltage bulb which still only draws 300mA. In other cases, a source is, by design, simply incapable of supplying a matched load, a good example being a 660MW turbo-alternator. With this, the design minimum value of the load is about 30 or more times the internal resistance – overload protection devices would trip long before the matched load condition were met.

In the design phase of electronic modules where matching is important, such as TV camera signal processing chains, telephony cable repeaters, radio receivers and transmitters of all sorts, extensive use is made of test equipment such as signal generators, spectrum analysers and the like. The sources are designed to produce an accurately known output level, such as -10dBm, into a matched load. -10dBm means a level of -10dB relative to a power of 1mW delivered to a matched load, or 100µW. In telephony, where a 600Ω impedance system is common, 1mW corresponds to

0.775Vrms, and telephone engineers often define 0dBm as meaning 1mW in 600Ω. But the more common usage is to define it as 1mW in whatever the system design impedance is – corresponding to 225mV in a 50Ω system (common in RF equipment), 273mV in 75Ω (common in TV baseband signal working) or 387mV in 150Ω (twisted pairs in underground cables).

In radio frequency testing, a module's input port is commonly driven by a signal generator with a purely resistive output impedance of 50Ω, and its output port terminated in the 50Ω resistive input impedance of a spectrum analyser. In the case of the module's input, the power delivered to it will be very close to that which would be delivered to a 50Ω resistor, even if the module's input impedance departs fairly markedly from 50Ω resistive. This is illustrated in Figure 29.1(a) and (b), where things have been normalised to unity, i.e. a generator with a 1Ω source delivering 1W to a nominal 1Ω load. For reasons explained in the Box, provided $R_S = 1\Omega$, the power in the load is close to 1W even if R_L varies – even if R_L is one third of an ohm, or 3Ω, the power in the load is 750mW, or only -1.25dB for a VSWR of 3 : 1! However, although the power in the load is not very sensitive to variations in R_L, the power dissipated internally in the source, and hence the total power supplied by the 2V ideal generator, varies markedly. Figure 29.1(b) shows that the total power supplied by the 2V generator varies from 4W for a short-circuited load, down to zero when R_L equals infinity.

Figures 29.1(a) and (b) show the situation at 0Hz or DC, where the effect of any incidental reactive terms in R_S and R_L can be ignored. The maximum power theorem applies equally at AC, but with the added complication that one is in general dealing with impedances rather than pure resistances. This is shown in Figure 29.1(c), where inductive and capacitive components are shown in R_S and R_L respectively, though it could equally well be the other way round, or both reactances could be of the same sign – positive for inductances and negative for reactances. In Figure 29.1(c), if at the frequency of the sinewave source E_{source}, the reactance of the inductive component L_S of the source equals that of the capacitive component C_L of the load, then they cancel each other out and the power in the load is determined purely by the values of R_S and R_L. This is known as the conjugate matched condition and it can only occur when the reactive components of R_S and R_L are of opposite signs. Since, if the frequency of the E_{source} signal increases, the reactance of L_S increases whilst that of C_L decreases, the conjugate matched condition can only occur at the one frequency – conjugate matching is inherently narrow band. For this reason, signal generators are designed such that Z_S is as nearly as possible purely resistive, any L_S or C_S being ideally zero. A similar comment applies to the input impedance of measuring instruments, such as power meters and RF spectrum analysers.

So if Z_S in Figure 29.1(c) is purely resistive and equal to the system design impedance, the power in the load is relatively independent of its exact

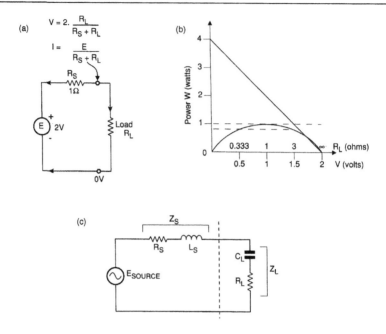

Figure 29.1 *(a) Normalised source and load, illustrating the maximum power theorem. (b) Power in the load resistor R_L as a function of its value, when R_S equals the design system impedance of 1Ω (curve) and total power supplied by the generator (sloping line). (c) In the AC case, source and load reactance must be taken into account. Source to the left of the dotted line, load to the right.*

value. But what if the dotted line in Figure 29.1(c) had been drawn horizontally, making what is now Z_S the load – exactly 1Ω, say – and a variable Z_1 now the source resistance? Considering the DC case, the power in a fixed load of 1Ω as R_S varies from zero to infinity is now given, in Figure 29.1(b), by the vertical distance between the curve and the sloping line: if R_S varies, the power in the load varies wildly, even if the load is a pure resistance equal to the system design impedance. Is this important? The following cautionary tale shows that it is.

Some years ago, the company where I then worked was developing an advanced all-band surveillance receiver of modular construction; a project in which as it happened I was not involved. One engineer designed the front-end half-octave filter module, another the RF amplifier and first mixer module, another the first IF and so on through the second and third mixers and IFs, the design aim being that on servicing by module replacement, replacing any or all module(s) should leave the overall performance within specification. To this end, a VSWR tolerance was placed on the input and output impedance of each module. As development progressed, the insertion gain of each module (or insertion loss in the case of the front-end filters) was checked using a signal generator and spectrum analyser, or a network analyser as available, the latter also checking port impedances. Thus module inputs were driven from a respectable 50Ω source, and their output checked with a faultless 50Ω load. Nevertheless, complete receivers exhibited a range of performance which was outside the specification limits. For unlike the situation on test, where each module port connected to a 50Ω interface, at the interface between modules in use, both port impedances could be different from 50Ω.

This effect is (or should have been) well known, graphs documenting it having appeared in one issue of the Marconi house magazine many years ago, but a search through my files failed to unearth it. So I worked it out again, for the simplified case where both R_S and R_L vary, but both are resistive. Figure 29.2 shows how the power in the load R_L varies with the value of R_L, for a series of different values of R_S from 0.25 to 2Ω, the source emf behind R_S being 2V as in Figure 29.1(a), call this Case I. When R_S equals 1Ω, maximum power naturally results in a matched 1Ω load, this curve being the same as in Figure 29.1(b). And in accordance with the maximum power theorem, when R_S equals 0.25Ω – top curve – maximum power, 4W, occurs in the load when its value is also 0.25Ω. To a nominal 1Ω load, $R_S = 0.25\Omega$ delivers 2.56W or +4.1dB, and significantly, the power changes rapidly for small deviations of R_L from unity. Likewise, when R_S equals 2Ω, maximum power, 0.5W, occurs in a 2Ω load, while when $R_L = 1\Omega$ the output is -3.5dbW.

The above case applies where the source emf is what it should be – twice the voltage across a matched load, but the source resistance R_S is incorrect. This corresponds to the case of a signal generator where the output

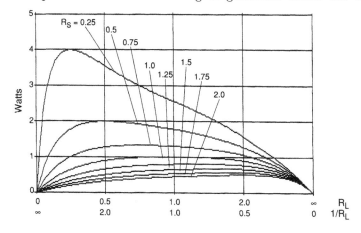

Figure 29.2 *Power in the load resistor R_L as a function of its value, showing curves for several different values of R_S. Source emf E_{source} as in Figure 29.1(a).*

impedance-defining resistor is damaged, but the instrument is otherwise unchanged from new, perhaps a rather unusual case, or perhaps due to the inadvertent application of RF power to the signal generator's output. A different set of circumstances arises in Case II where a module with a poor output impedance (perhaps a synthesiser used in a receiver) has been set up to deliver its rated output to a resistive load equal to the system design impedance, e.g. 50Ω. In this case, the internal Emf E_{source} will have been effectively adjusted (speaking in terms of the normalised circuit of Figure 29.1) to something other than 2V, so as to deliver 1W into a 1Ω load. Thus if R_S is lower than 1Ω, E_{source} will have been set to less than 2V, and to more than 2V if higher. The power delivered to the load R_L, as a function of the value of R_L, for various values of R_S, is shown in Figure 29.3. Since E_{source} has been adjusted to deliver 1W into 1Ω whatever the value of R_S happens to be, all the curves pass through 1W at $R_L = 1\Omega$. But only in the case where $R_S = 1\Omega$ is the curve horizontal at $R_L = 1\Omega$, giving the relative independence of load power versus R_L that obtains when the value of R_S is correctly set at the nominal system impedance. Nevertheless, the variations of power delivered with variation of both R_S and R_L are much less in this case than Case I, permitting the results for Case II to be plotted in Figure 29.3 with a vertical scale twice that of Figure 29.2.

Correct matching is particularly important where filters are concerned, as the following cautionary tale illustrates. Sometime in the 1980s, my then boss came up to me concerning a crisis with the company's new HF receiver, which was already over budget and overdue for delivery. The 20–30MHz sub-octave filter was far too narrow, and excessively lossy to boot. 'It's got to

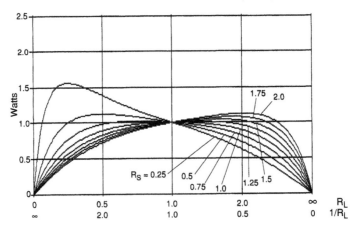

Figure 29.3 *Power in the load resistor R_L as a function of its value, showing curves for several different values of R_S. Source emf E_{source} adjusted to give 1W into a 1Ω load resistor R_L, for each value of R_S.*

be fixed, but don't spend more than a week on it, even if the conclusion is that we have to go for a spec. relaxation.' Having spent the rest of the morning getting the necessary test equipment together, I was able to report half way through the afternoon that the filter was now working fine. So it should, for it was a seven-pole elliptic design straight out of Ref. 1, and checking the values on the circuit diagram confirmed that the engineer who had designed it had done his denormalising sums right. The capacitors on the board were also all correct, and the coils all capable of being tuned by means of their slugs to the correct inductance. The output of the filter module was normally connected to the RF/first mixer module, which presented a good low VSWR input, but before getting there, the output of the 20–30MHz filter, when selected, had to pass through a number of band-select relays and board tracking, which looked distinctly capacitive. As far as the filter was concerned, this capacitance was part of the load, which should have been purely resistive but wasn't. Reducing the value of the final shunt capacitor in the filter effected an improvement, and further reduction made it better still. In the end, it turned out that no capacitor was needed at all, the circuit strays equalling the design value of the filter's final shunt capacitor. But in the end I settled for 1.8pF, to avoid C_{99} on the circuit diagram being shown as 0pF. The final capacitor in the 15–20MHz was also reduced in value somewhat, the lower frequency filters being in spec. due to the much larger values of their final capacitors.

Box

If R_L in Figure 29.1(a) increases by 1% to 1.01Ω, then the total circuit resistance (R_S being 1Ω) increases by 0.5%. Thus the current *decreases* by 0.5%. The power P dissipated in R_L is given by $P = i^2 R_L$. If i decreases by 0.5%, then i^2 decreases by 1%, but R_L has increased by 1%, so the product is (virtually, to a first order) unchanged. This is a result of the Binomial Theorem, but can equally be verified by working out $i^2 R_L = (2/(1+1.01))^2 \times 1.01$ on a pocket calculator.

Some results from the Binomial Theorem:

$$(1 + \delta)^2 = 1 + 2\delta \qquad (1 - \delta)^2 = 1 - 2\delta$$
$$(1 + \delta)^{-1} = 1 - \delta \qquad (1 - \delta)^{-1} = 1 + \delta$$
$$(1 + \delta)^n = 1 + n\delta \qquad (1 - \delta)^n = 1 - n\delta$$
$$(1 + \delta)^{-n} = 1 - n\delta \qquad (1 - \delta)^{-n} = 1 + n\delta$$

Note: These results only apply if $\delta << 1$ and n is smallish, so that second-order and higher terms are insignificant.

Reference

1. Geffe, P. R. *Simplified Modern Filter Design*, Iliffe Books Ltd, London 1964.

Index

Printed and bound by CPI Group (UK) Ltd, Croydon, CR0 4YY

03/10/2024

01040435-0015